S0-EID-520

NATIONAL GEOGRAPHIC KIDS

NERDlet[1.]

1. (n.): a little book of nerdy stuff

1 2 3 4 5 6 7 8 9 10

TECH

Jamie Kiffel-Alcheh

NATIONAL GEOGRAPHIC
WASHINGTON, D.C.

BUG BOTS, PAGE 112

INTRODUCTION

ROBOTS. VOICE COMMANDS. LASERS.

If these words give you goose bumps of excitement, read on! This book is fully programmed with the coolest facts about all things tech. You'll find funny stuff, quirky stuff, weird stuff, and just plain awesome stuff. LOL tech? Check. Tech personality quizzes? Check. Techie tidbits you can spring on your friends to knock them out with your super smarts? Definitely check.

So, if bits and bytes, spy cams, face-changing apps, and cyborgs are things you'd like to know a lot better, come on in. Three ... two ... one ... warp speed! We're launching into the tech universe and beyond.

EVTOLS, PAGE 82

LASERS, PAGE 6

HUG-A-BOTS,
PAGE 166

NATIONAL IGNITION FACILITY

FUN FACT

LASER is the acronym for "light amplification by stimulated emission of radiation."

YOU MAY ALREADY KNOW LASERS CAN CATCH SPIES, slice through metal, and drive cats crazy. But what happens when lasers get really big? The National Ignition Facility (NIF) in California, U.S.A., is 10 stories tall and as wide as three American football fields, making it the biggest laser ever. And what it does is literally out of this world: Scientists focus all of NIF's 192 laser beams at once onto a pencil eraser–size target filled with a tiny amount of the radioactive elements deuterium and tritium. This raises the target's temperature to more than 180 million degrees Fahrenheit (100 million °C)—several times hotter than the sun's center—and creates 100 billion times more pressure than Earth's atmosphere. The result: a miniature star! OK, so the star only lasts for 100 trillionths of a second, but while it does exist, it joins atoms together in a process called nuclear fusion. Zap!

NERD ALERT: *SCI-FI COOL*

NIF IS THE MOST PRECISE laser on Earth and the world's biggest optical instrument. And, when it creates a miniature star, it temporarily generates one of the hottest spots in our solar system! Even better: Scientists hope to someday use nuclear fusion to create eco-friendly energy.

FUN FACT

The first lasers ever made tapped out at 10,000 watts—about as powerful as 100 standard lightbulbs. Today, NIF packs more than 500 trillion watts.

VOICE COMMAND

AMAZON'S VOICE ASSISTANT, ALEXA, CAN ANSWER YOUR QUESTIONS, PLAY YOUR FAVORITE SONGS, AND MORE.

SIRI, ALEXA, AND GOOGLE ASSISTANT can seem almost human when they answer your questions. But how do these voice assistants "know" what you're asking? That depends on the device. The simplest systems recognize words by comparing them with prerecorded phrases in their memory banks. More sophisticated ones break down your words into separate sounds, then ID them by the unique sound waves they produce. But if you say "Which way is Botswana?" the computer needs to figure out whether you mean "way" or "weigh." Only the most advanced devices can do this, and they work by using probability. Statistically speaking, it's unlikely that your sentence includes "weigh." So, the computer concludes you mean "way." When it messes up, remember how tricky it is for a human to learn the differences between "there," "they're," and "their," and you might forgive your digital assistant.

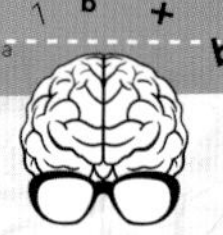

NERD ALERT: SUPER SMARTS

EACH HUMAN LANGUAGE IS MADE UP of sound building blocks called phonemes. There are about 44 of them in English, and advanced voice assistants "learn" them all!

FUN FACT

The most popular cooking question for Alexa is "How do I boil an egg?"

FUN FACT

More than half a million people have told Alexa "I love you."

TECH GONE WILD

EVEN THE BEST ENGINEERS CAN'T OUTMATCH MOTHER NATURE. CHECK OUT THESE UNTAMED TECH BREAKTHROUGHS—ALL INSPIRED BY ANIMALS.

GECKO GRIPPERS

TAPE AND GLUE both lose stickiness after they're used. But geckos' feet stay sticky, thanks to tiny hairs that cause electromagnetic attraction. That's why NASA engineers copied the hairs to make grippers. These may one day allow robots to climb outside the International Space Station and repair it ... without floating away.

ROBO-TRUNK

FOR YEARS, robotic arms used in factories were modeled after human arms. But those arms were hard and rigid, and if human workers got in their way, people could be hurt. The Bionic Handling Assistant is bendy like an elephant's trunk. It's made of a series of gray plastic disks with air between them, so it's both softer and safer than traditional robotic arms.

WHALE POWER

A TURBINE IS A KIND of propeller that generates energy as it pushes through wind or water. For a long time, researchers didn't realize a bumpy turbine blade would be more efficient than a smooth one. But after studying and copying the bumpy flippers of humpback whales, scientists discovered these bumps can help submarines and planes dive and turn more acrobatically—like whales do.

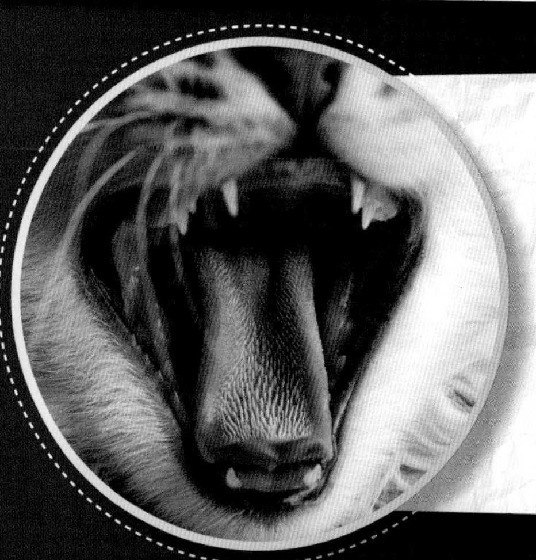

CAT TONGUE HAIRBRUSH

IF YOU'VE EVER BEEN LICKED BY A CAT, you know its tongue feels like sandpaper. That's from thousands of miniature, flexible spines called papillae on the cat's tongue. An artificial version that looks like a large rectangular-shaped plastic cat tongue with a handle can undo knots better than a standard hairbrush. It might also work as a special cat brush that can remove allergy-causing dander and help vets apply skin medicines to felines.

ROBO HIGH

CLASS SUPERLATIVES

Big Bot on Campus

WITH A NAME MEANING "SAMURAI WARRIOR" and a sky-scraping height of 28 feet (8.5 m)—that's like one giraffe standing on another giraffe's head—MONONOFU is the world's largest humanoid vehicle. Built in Japan, the bot is officially a vehicle because, to make it walk, a human operator has to sit inside its cockpit head. How tall is that head? To get there, you'll need to take its built-in lift, which is a mechanical harness that hoists you to the top! It isn't easy to see the ground from MONONOFU's small cockpit, so five internal monitors help keep you from stomping your friends. Maybe it's for the best that MONONOFU's too tall to get out of its warehouse without disassembly.

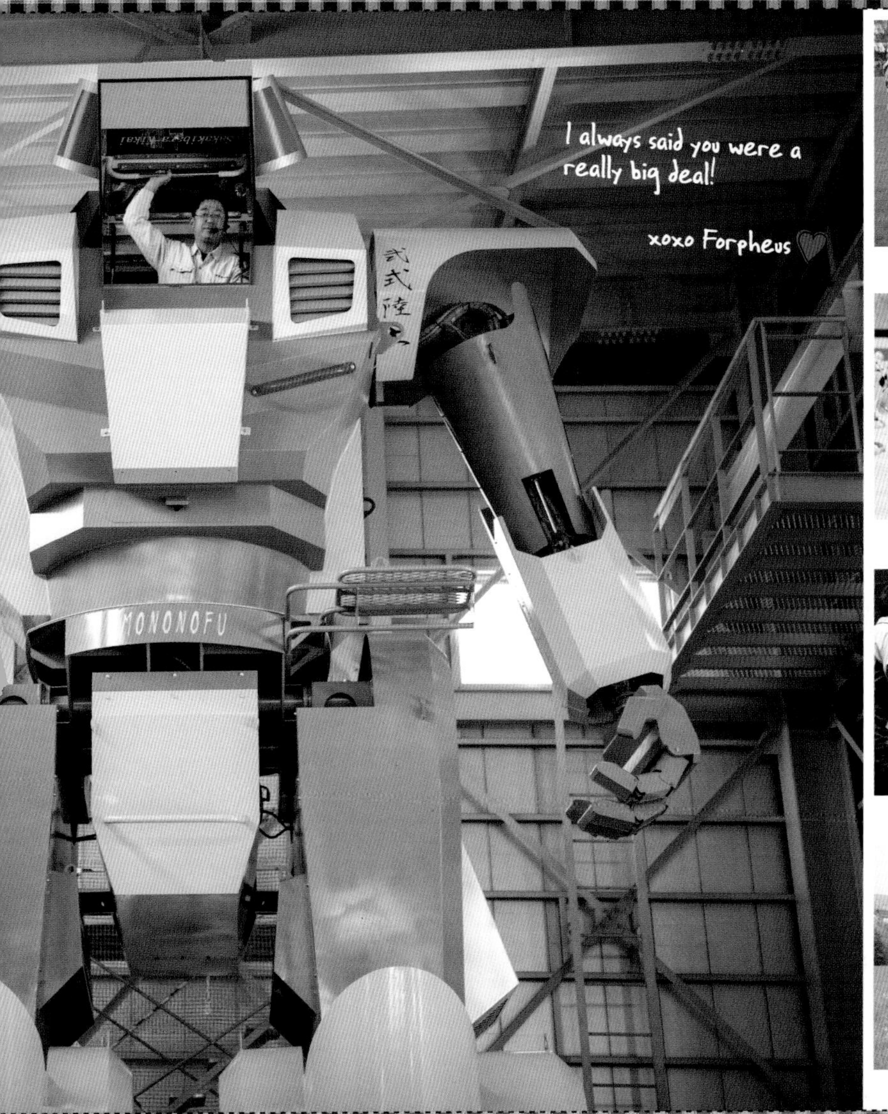

Xingzhe No. 1, **page 56**

Alpha 1S, **page 96**

Forpheus, **page 138**

Tradinno, **page 174**

SPACE SUITS

ASTRONAUT ANNE MCCLAIN SUITS UP FOR TRAINING.

THE STYLES YOU CHOOSE TO WEAR ON EARTH are up to you, but in space, there are strict fashion rules. For instance, your duds have to keep you warm when it's minus 250°F (-157°C), and cool you off if conditions heat up to 250°F (121°C). NASA's space-walking suits can do all that and a lot more. Officially speaking, they're one-person space-crafts. They supply oxygen through a backpack and contain drinking water. Each is strong enough to deflect space dust, which moves faster than a bullet, and its visor protects against searing sunlight, which is even brighter in space than on Earth, where our atmosphere scatters and absorbs part of it. In case of an emergency, jets in a suit's back can propel the astronaut to safety. Move over, Buzz Lightyear!

NERD ALERT: SCI-FI COOL

SPACE SUITS ARE BEING REDESIGNED to allow more people than ever to go to the moon. Unlike past suits, the new xEMU suit, made especially for moonwalks, comes in nearly every person's size! It's also much more flexible than past lunar suits, so astronauts can walk and not just "bunny hop" along the surface.

THE XEMU SUIT IS MORE FLEXIBLE THAN TRADITIONAL SPACE SUITS.

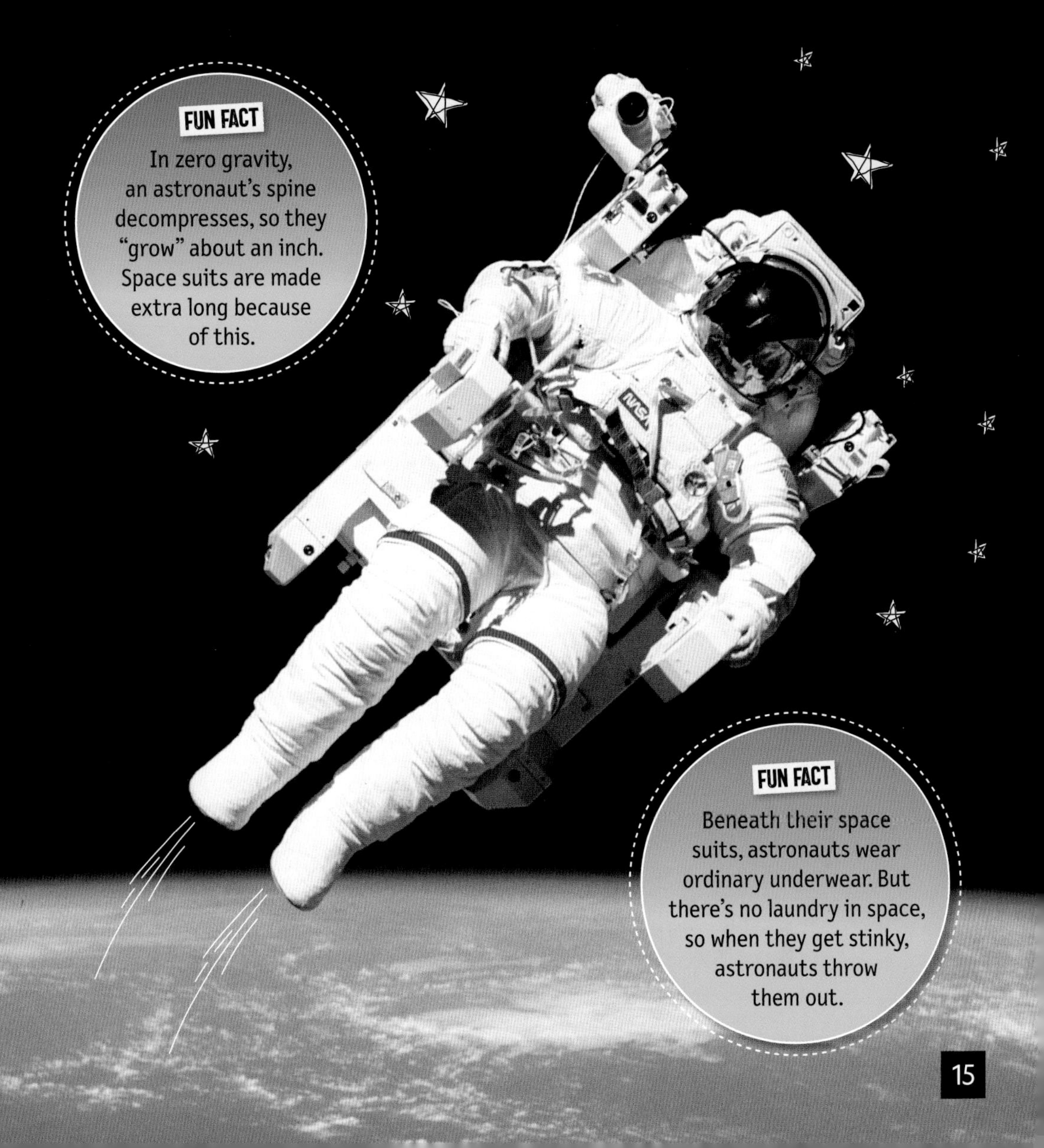

FUN FACT

In zero gravity, an astronaut's spine decompresses, so they "grow" about an inch. Space suits are made extra long because of this.

FUN FACT

Beneath their space suits, astronauts wear ordinary underwear. But there's no laundry in space, so when they get stinky, astronauts throw them out.

FACE-CHANGING APPS

YOU UPLOAD A PHOTO, press a button and ... turn into your grandmother?! The secret behind this transformation is a type of artificial intelligence known as an artificial neural network (ANN). An ANN is meant to imitate real neural networks—that is, human brains—by, well, learning. ANNs are initially programmed with a training set of images. Each time an ANN makes a correct match between a new pic it's given and its training set, it learns. So the more 85-year-olds with wrinkles the ANN "sees," for instance, the more easily it recognizes, and also copies, how an average 85-year-old looks. By noting the exact positions of bags and sags, the ANN uses that info to adjust your pic and make it look older. ANNs can also change hairdos, let you swap faces, or turn you into other creatures!

THANKS TO APPS, YOU CAN TRY OUT A NEW HAIR STYLE BEFORE COMMITTING TO THE CHOP.

FUN FACT

ANNs were invented by a psychologist in 1958. The first one was made to model how our brains process what we see. That ANN learned to recognize objects!

NERD ALERT:
SUPER SMARTS

IT'S NOT ALL FUN AND GAMES: Because they process gigantic data sets ultra fast, ANNs sometimes solve problems humans can't. Researchers designed an ANN that may soon answer gravity questions astronomers have been puzzling over since the time of Sir Isaac Newton.

WHAT ELSE CAN ARTIFICIAL NEURAL NETWORKS DO? GET THE FAR-OUT DETAILS FROM THE NERD OF NOTE ON THE NEXT PAGE.

NERD OF NOTE:
AMBER YANG

Amber was always interested in space. While watching livestreams of astronaut Scott Kelly, she noticed something: Kelly had to exit the International Space Station (ISS), taking cover in a separate capsule, when the spacecraft was in danger of being struck by flying debris. Space junk is no joke—this stuff moves at 22,000 miles an hour (35,400 km/h)! The junk can be anything from the glove that floated away during the first spacewalk to entire discarded rocket boosters.

Amber learned that there were space debris surveillance networks that helped predict where debris would go, but they weren't reliable. Weather and atmospheric changes threw off their models. Plus, they had to be manually updated constantly. That got her thinking: Could artificial intelligence (AI) learn to recognize the debris' orbit patterns and predict future orbits?

Though she'd never worked with AI before and says she barely knew how to code, Amber learned as she went. She tweaked her project for her high school science fair, then realized the software was good enough to take seriously. Today, her company, Seer Tracking, studies past positions of space debris and detects patterns in how it moves. From these data, it predicts the space debris' position and velocity—in other words, where it will turn up next.

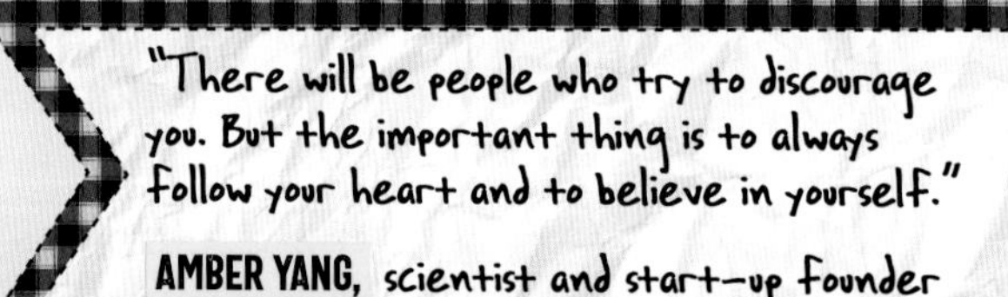

"There will be people who try to discourage you. But the important thing is to always follow your heart and to believe in yourself."

AMBER YANG, scientist and start-up founder

Space debris buildup on the ISS

CAMERA TRAPS

A CAMERA TRAP SNAPS A PIC OF CURIOUS TIGER CUBS IN INDIA.

A DUNG BALL SLOWLY ROLLS past a wild elephant. To the elephant, it probably seems like a totally normal, boring dung ball. But the rolling poo actually contains a Wi-Fi–enabled camera! The dung is a camera trap—a tool scientists use to snap animals' pics without their noticing. Some camera traps, like the dung cam, are controlled remotely. Others are triggered by an animal's motion or body heat. These tools give researchers and photographers the chance to get shots of creatures that wouldn't otherwise stick around to be photographed, or are endangered and hard to find.

NERD ALERT: DO-GOOD GEAR

SOME OF THE MOST VALUABLE SHOTS DON'T STAR ANIMALS AT ALL. They catch poachers. In Asia, when hidden cams catch pics of would-be tiger poachers, they send them straight to forest rangers, who jump into action. Some of the latest camera traps, created by wild cat conservation group Panthera, use artificial intelligence to recognize humans in images so that poachers can be nabbed on the spot.

FUN FACT

Camera traps help researchers count how many endangered tigers live in a single area by capturing their stripe patterns; each tiger's pattern is unique.

FUN FACT

Some researchers have built robotic animal "spies" with cameras for eyes.

A DUNG-DISGUISED CAMERA CALLED PLOPCAM HELPS BBC FILMMAKERS CAPTURE ELEPHANT FOOTAGE.

WHAT'S YOUR RETRO ARCADE GAME?

THERE ARE SOME AWESOME RETRO GAMES, BUT WHICH WOULD BE YOUR GO-TO? FIND OUT!

START HERE.

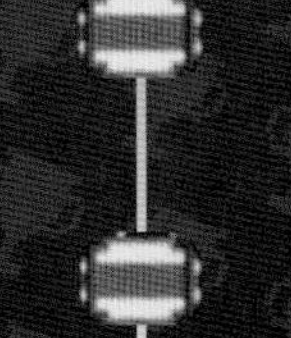

ARE YOU GOOD AT MAZES?

Yes!

WHAT WOULD YOU RATHER EXPLORE—A WILD JUNGLE OR A MYSTERIOUS CAVE?

Cave

DIG DUG
Your game is Dig Dug! You'll dig mazes to escape enemies.

Jungle

PITFALL
Your game is Pitfall! Navigating a maze of jungles on a timer is just for you.

Can I have both?

I'm more into strategy games like chess, or at least tic-tac-toe.

ARE YOU A MARATHONER LIKE A PRONGHORN, OR A SPRINTER LIKE A CHEETAH?

I'm a sprinter — short bursts are my strength.

PAC-MAN
Your game is Pac-Man or Ms. Pac-Man! You'll chomp your way to success.

I can run marathons.

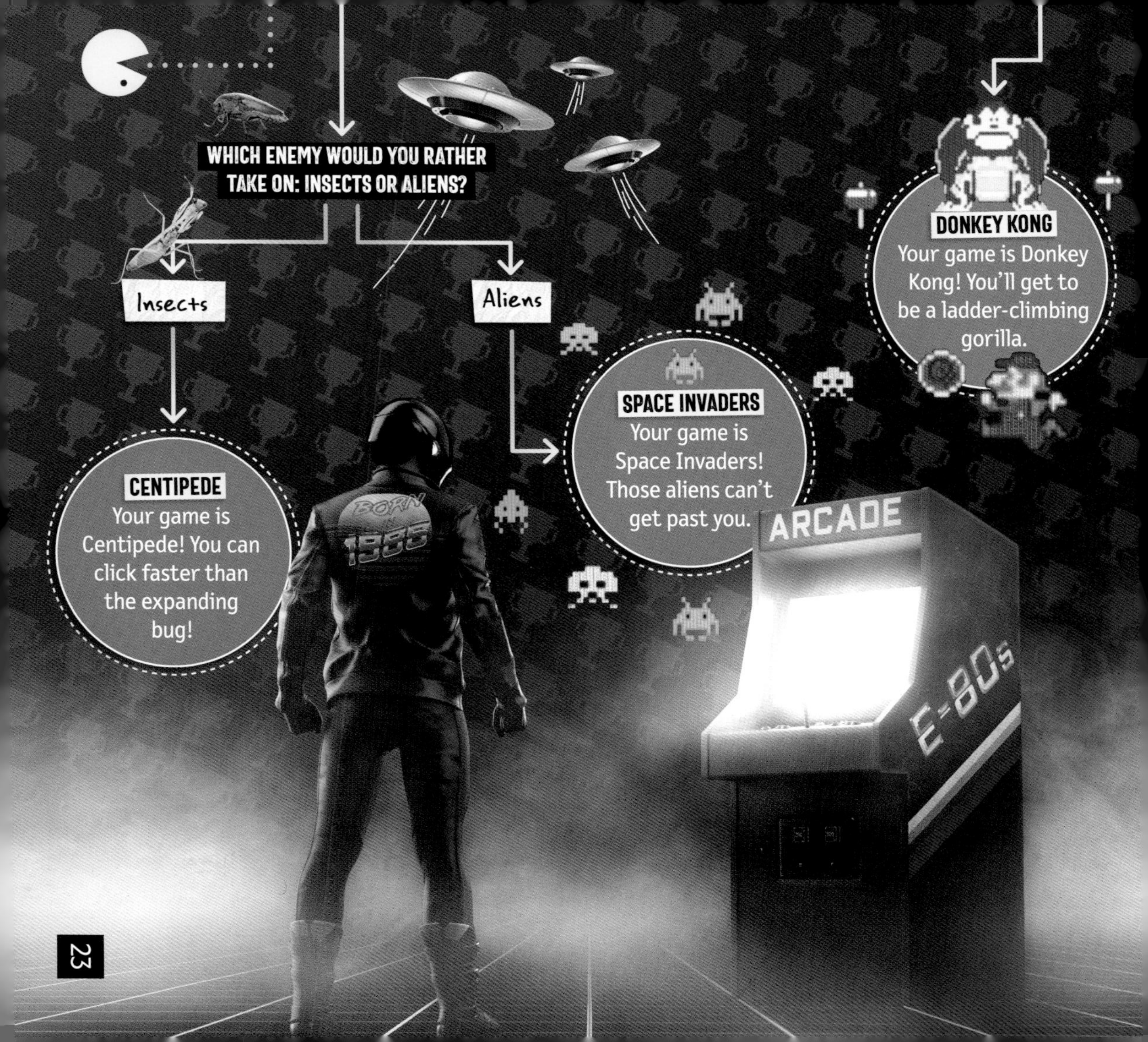

WHICH ENEMY WOULD YOU RATHER TAKE ON: INSECTS OR ALIENS?

Insects

Aliens

CENTIPEDE
Your game is Centipede! You can click faster than the expanding bug!

SPACE INVADERS
Your game is Space Invaders! Those aliens can't get past you.

DONKEY KONG
Your game is Donkey Kong! You'll get to be a ladder-climbing gorilla.

ROBONAUT

IT LOOKS LIKE A MEMBER OF DAFT PUNK with Gumby's body, but it can't sing. That doesn't mean you should underestimate Robonaut! This high-tech bot has an important job: It was designed by NASA to assist humans in space. With "climbing legs" that grip and hang from a railing, fingers nimble enough to flip a switch or lift a flap, and the ability to follow conversational commands, Robonaut was built to take over the jobs humans cannot (or would rather not) do. If a task is dangerous or repetitive, Robonaut can step in. One day, it could even become a scout to explore Mars.

FUN FACT

Robonaut's first tasks included handrail cleaning—not too glamorous, but a major time-saver for the human crew.

NERD ALERT: SCI-FI COOL

ROBONAUT'S SENSORS AND VISION capabilities mean it can use a keypad and identify different buttons. And get this: Because human legs aren't very useful in space (they're mostly used for floating), Robonaut's legs are made to function like an extra pair of arms. It can use them to work while "standing" right side up or upside down.

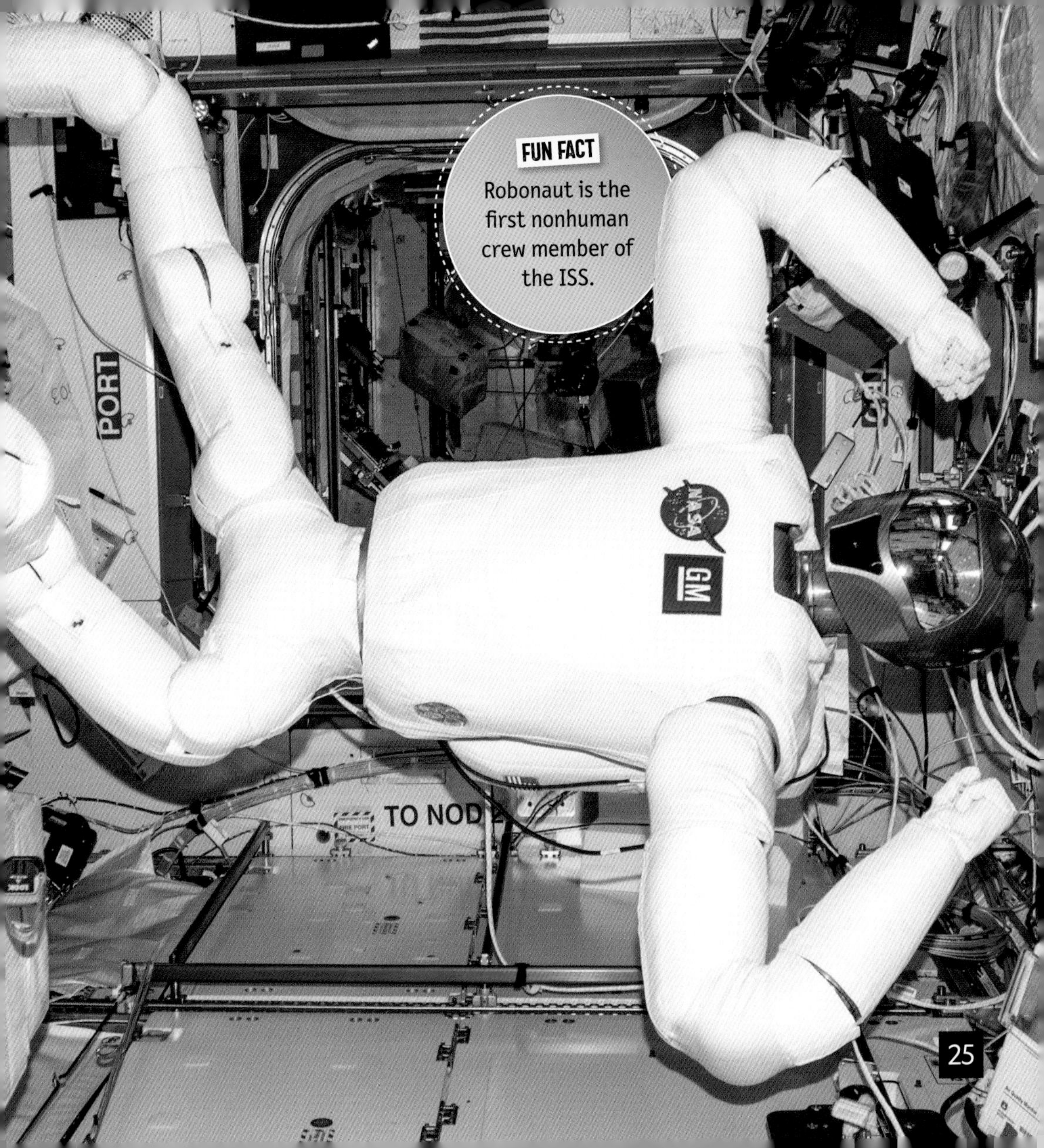

FUN FACT

Robonaut is the first nonhuman crew member of the ISS.

CONSTRUCTIVE IDEAS

TECH MAY SOON SUPERCHARGE BUILDINGS ALL OVER THE WORLD. HERE ARE SOME OF THE LATEST CONCEPTS AND CREATIONS.

EARTHQUAKE SPRAY

MANY STRUCTURES AREN'T made to stand up to earthquakes, and in some parts of the world, reinforcing them is too expensive. Enter spray-on concrete reinforcement. Working like a flexible, protective layer over fragile concrete, it can help a building stand up to a magnitude 9.1 earthquake by letting it bend without breaking during the quake.

TERMITE-INSPIRED TRICKS

IN ZIMBABWE, the 350,000-square-foot (32,500-sq-m) Eastgate Centre uses 90 percent less energy than the similar size building next door, thanks to some construction tricks learned from bugs. The building has vents all over its top and sides that release heat and help circulate cool air, just like in termite mounds.

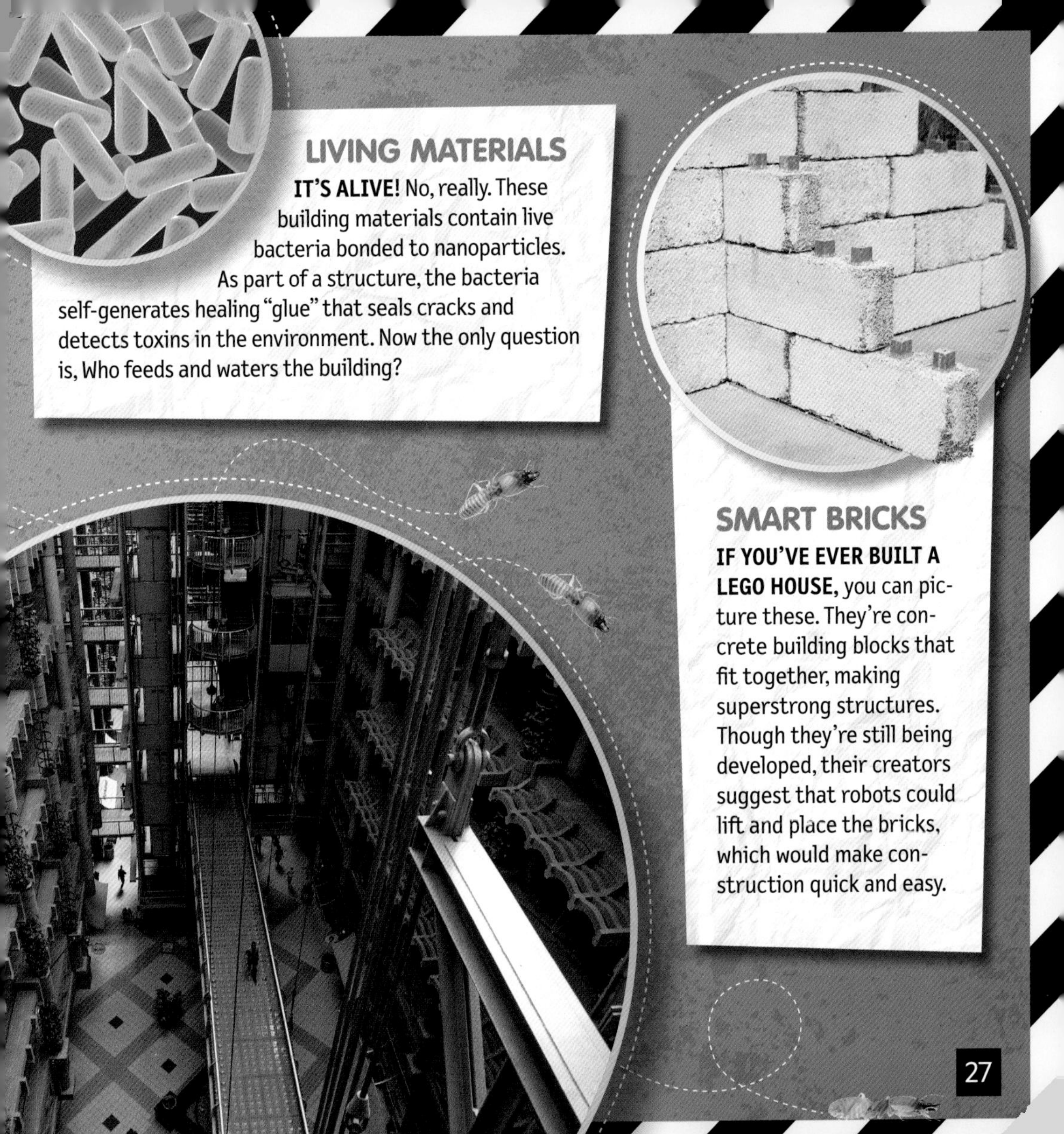

LIVING MATERIALS

IT'S ALIVE! No, really. These building materials contain live bacteria bonded to nanoparticles. As part of a structure, the bacteria self-generates healing "glue" that seals cracks and detects toxins in the environment. Now the only question is, Who feeds and waters the building?

SMART BRICKS

IF YOU'VE EVER BUILT A LEGO HOUSE, you can picture these. They're concrete building blocks that fit together, making superstrong structures. Though they're still being developed, their creators suggest that robots could lift and place the bricks, which would make construction quick and easy.

SMART ROADS

THIS BIKE PATH IN THE NETHERLANDS IS PAVED WITH SOLAR PANELS.

YOU'RE ALMOST LATE TO softball practice when—ping!—your car suggests a route with less traffic. That's not a navigation app talking. It's the road communicating through the car! "Smart roads" contain sensors that recognize how many vehicles are driving on them, their locations, and how fast they're going. Some smart roads, currently in development, may even be able to call for help when a crash happens. Others include photovoltaic panels (solar panels) that store power by day and light highways and surrounding areas at night. These lights can save even more energy by turning on only when cars pass them. Engineers are also developing roads with temperature-sensitive, glowing paint that illuminates when it's icy, warning drivers to slow down before they slip or slide.

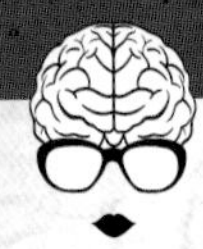

NERD ALERT: SUPER SMARTS

WITH THE RIGHT PROGRAMMING, smart roads will help not only people but also the planet. Several countries are looking into adding "piezoelectric energy harvesters" to roads. These devices contain crystalline structures that generate a charge when vehicles drive over them. Heavy traffic might make enough electricity to help power cities!

FUN FACT

Smart sensors aren't the only way roads are becoming better than ever: Some new, eco-friendly ways to make roads include reusing ground-up rubber, plastic trash, and food waste!

MACREBUR
The plastic road company
THIS ROAD IS MADE FROM
WASTE PLASTICS

FUN FACT

"Blacktop" isn't always black. In Arizona, U.S.A., you can find red roads made with local red rock.

THE OCEAN CLEANUP

HERE'S THE BAD NEWS: Every year, millions of tons of plastic wash into the oceans, then break down into microplastics—tiny, harmful pieces of litter that sea creatures ingest. The good news: People are working hard to clean up the mess, and using high-tech solutions to do it. In 2013, Dutch teen Boyan Slat came up with a system called The Ocean Cleanup, a floating barrier made to catch and recycle waste. The system is basically a giant net that traps any trash the current washes into it. Every six to eight weeks, a ship removes the plastic and takes it to be recycled. While the full system isn't yet in effect, its creator says that once it is, it will take just five years to clear out half of the Great Pacific Garbage Patch, one of the world's largest concentrations of plastic waste.

WORKERS PULL HUGE AMOUNTS OF GARBAGE—INCLUDING OLD NETS—OUT OF THE PACIFIC OCEAN.

NERD ALERT: DO-GOOD GEAR

THE OCEAN CLEANUP HAS A BIG, IMPORTANT JOB TO DO: The Great Pacific Garbage Patch is one of several massive patches of litter that swirl from the surface to the seafloor. It flows between Hawaii and California, U.S.A.—about 2,400 miles (3,862 km)—and is packed with more than 1.8 trillion pieces of plastic for the Cleanup to catch.

FUN FACT

Environmental activist Ben Lecomte swam through the Great Pacific Garbage Patch in 2019 to show the world how much debris is in the ocean.

FUN FACT

Plastic debris can be anything from single-use water bottles to any number of weird and wacky items. In 1992, tens of thousands of bathtub toys fell off a cargo ship, filling the Pacific with rubber duckies!

iMAGINE THIS

SUPER-NERDS!

WHAT'S BETTER THAN A SUPERHERO? HOW ABOUT A SUPERNERD? IF YOUR FAVORITE COMIC BOOK STARS TURNED ULTRA NERDY, THEY'D HAVE TO WIELD SOME WACKY NEW GADGETS.

HERO:

BATMAN

GADGET: Exoskeleton (robotic suit that supercharges the wearer's strength and endurance)

WHY HE LOVES IT: Unlike a lot of other superheroes, Batman is still a regular human and can tire out. So when the going gets tough, he can put on an exoskeleton, and its sensors and artificial intelligence will double his endurance. Ha, ha, Joker!—this joke's on you.

HERO:
WONDER WOMAN

GADGET: Lost and found tracker (digital item-finder that sticks to stuff you don't want to lose)

WHY SHE LOVES IT: Wonder Woman relies on her bracelets and shield to stop flying bullets. But what if an enemy swipes them? A lost and found tracker can ping her phone and alert the Justice League if her magical weapons get out of range.

HERO:
SUPERMAN

GADGET: Nanosponge (injectable, microscopic polymer sponge)

WHY HE LOVES IT: A nanosponge would soak up kryptonite toxins as it swooshed through Superman's bloodstream, safely wiping them out of his body. Take that, Lex Luthor!

LEAFY GREEN CITIES

A BAMBOO WALKWAY WINDS THROUGH A BAMBOO FOREST IN KYOTO, JAPAN.

WHEN YOU THINK OF THE URBAN JUNGLE, you're probably not picturing an actual jungle. But that might be exactly what future cities look like! "Vertical forests," which already tower above Milan, Italy, are apartment buildings covered in more than 20,000 plants. The greenery absorbs pollution and carbon dioxide, which could also reduce the effects of climate change. Meanwhile, architects worldwide are building cities using bamboo. Since bamboo is a grass, it grows back quickly after it's cut down—something trees can't do. It's also superstrong and can grow stalks harder than oak within five years. Some architects suggest covering bamboo buildings in plants, then surrounding them with bamboo forests that can be harvested and used to make the structures bigger and stronger.

NERD ALERT: DO-GOOD GEAR

BAMBOO'S SECRET, EARTH-FRIENDLY POWER is that it's sustainable, which means it won't quickly run out. Why not? It holds the record as world's fastest-growing plant. Certain species can grow up to about 36 inches (91 cm) per day. That's almost fast enough to see them growing!

FUN FACT

To prune the trees in a vertical forest, arborists hang from up to 361 feet (110 m) above the ground—that's as high as a football field standing on end!

FUN FACT

If the same number of trees in two vertical forest skyscrapers were in a natural forest, they'd cover more than 320,000 square feet (30,000 sq m)—about the size of eight supermarkets!

HOVERBOARDS

THE MOST COMMON HOVERBOARDS DON'T "HOVER," BUT THEY STILL FEATURE SOME PRETTY COOL TECH!

YOU MAY HAVE SEEN THEM WHIZ BY, or maybe you've zoomed around on one yourself. But do you know how hoverboards (also called self-balancing scooters) work? As sci-fi as they look on the outside, what's inside is even cooler: a central processing unit (CPU) that works a lot like an electronic brain. This CPU gets info from circuits—called microprocessors—all around the hoverboard. So when you shift your weight in any direction, a microprocessor picks up the message and sends it to the CPU. A gyroscope (a 360-degree spinning wheel) inside the board helps detect exactly where the board is leaning and makes the wheels spin in that direction. So where are the motors? There's one inside each wheel, and they generate the power to propel you. Lean in for speed!

A HOVERBOARD BALANCES YOUR BODY AND BALANCES COMPLEX INFO. Inside the board are switches with tiny LEDs and infrared sensors. As long as the sensors "see" the LEDs, the board stays still. But shift your weight forward, and you'll be pressing down the front switch. This makes a plastic tab slide down and block the LED. No light tells the board "Go!"

FUN FACT

The word "hover-board" was first used in the sci-fi comedy *Back to the Future II*, released in 1989.

FUN FACT

You're likely to be better at hoverboarding than your parents. That's because humans lose some balance starting around the age of 25.

DOES READING ABOUT HOVERING MAKE YOU WANT TO FLY? LEARN HOW TO CATCH SOME AIR ON THE NEXT PAGE!

FAQ WILL HOVERBOARDS EVER HOVER?

Wait a minute: Self-balancing scooters are great and all, but a real hoverboard would actually hover, right? It would float above the ground, taking you wherever you wanted to go, kind of like a magic carpet ... if that carpet were a skateboard. Well, the wait for real hoverboards may already be over. More than one company has created floating boards. Aerospace company ARCA has ArcaBoard, which resembles a giant wafer. It's got 36 fans to lift it one foot (0.3 m) off the ground and take you flying at about 12.5 miles an hour (20 km/h) for up to six minutes. Another hoverboard, the Flyboard Air, has soared 490 feet (150 m) in the air at a superfast 87 miles an hour (140 km/h), running on kerosene-fueled turbine engines. And the HoverBoard by ZR is a surfboard that attaches to a motorboat. It uses water power to send you flying as fast as 15 miles an hour (25 km/h). The downside: All these boards are expensive. The HoverBoard by ZR costs more than $5,000, and you'd need a boat, too. ArcaBoard costs about $20,000, and other hoverboards are similarly priced. So, odds are most of us will be waiting for our techie magic carpets a little while longer. But, hey, there is good news: An economy airline flight can take you much farther for a lot less!

FUN FACT

Carmaker Lexus built a hoverboard prototype that worked on superconductors cooled by liquid nitrogen. But it required a custom, magnetized skate park.

FUN FACT

Canadian software engineer Alexandru Duru built the first iteration of his Omni hoverboard using wood, propellers, and a remote control made from pliers!

FRENCH INVENTOR FRANKY ZAPATA ON THE JET-POWERED HOVERBOARD FLYBOARD AIR

FUN FACT

A still-in-development hoverboard called Hendo uses a magnetic field to make it levitate about an inch (2.5 cm) off the ground.

EXOSKELETON

WITH YOUR POWERFUL chest plate and motorized legs and arms, you look and feel like a Transformer. You step up to a 200-pound (91-kg) crate, take a deep breath, and pick it up. It's no sweat: You're wearing a Guardian XO exoskeleton, which gives you the strength you need to lift the heavy weight. The tech works by combining sensors with robotics, so when you move, motors kick in to move mechanized muscles. Some suits respond to your movement, while others pick up impulses from your brain. Exoskeletons can be used by soldiers, factory workers, and construction workers. They've also helped some people with disabilities to stand and walk.

A FACTORY WORKER USES HER EXOSKELETON TO CARRY HEAVY ITEMS.

NERD ALERT: DO-GOOD GEAR

EXOSKELETONS CAN BE POWERFUL TOOLS FOR SOME PEOPLE WITH DISABILITIES. Juliano Pinto, who has paraplegia, became the first person in an exoskeleton to perform the soccer World Cup kickoff. In 2014, while wearing an exoskeleton with a sensor-filled cap, he thought *kick*, and his robotically enhanced legs did just that!

FUN FACT

The world's first exoskeletons were made by ... nature! A crab's shell and an insect's hard body are exoskeletons. These creatures don't have a single bone in their bodies. After all, the word means "outer skeleton."

FUN FACT

About a decade ago, it took 6,000 watts to run an exoskeleton. Today, it takes less than 400—that's about as much as four fluorescent lightbulbs.

NERDSVILLE CENTRAL: AKIHABARA DISTRICT, TOKYO

EXPLORE ELECTRIC TOWN

IF TRANSISTORS, LEDS, CIRCUIT BOARDS, AND CAPACITORS CALL YOUR NAME, you'll want to put Akihabara on your travel list. Nicknamed Electric Town, it's basically the electronics mecca of the world. This shopping district is chock-full of specialty components, odd-size batteries, phones, computers, and used electronic parts—everything you need to build the tech nerd project of your dreams. There are megastores, shops, and stalls lining the streets. If you're more into playing with tech than making it, Akihabara is also a gold mine of gamer goods and anime stuff aimed at *otaku* (that's "superfans" in Japanese). For a dose of culture, you can visit the 1,200-year-old Kanda Myojin Shrine, where people go to get tech-related projects blessed. There's even an official *Omamori*, or traditional protective amulet, that visitors can buy to protect their digital info and devices.

1 Go on a weekend. There's a semisecret flea market below Akihabara Station's elevated railway tracks.

2 Follow the main street. Chuo Dori is the place where you'll find the most electronics shops and stands.

TIPS TO NERD OUT IN AKIHABARA:

3 Step into a SEGA arcade. You can play all the latest video games.

4 Visit a gachapon hall. These contain rows and rows of toy-vending machines. They're stocked with everything from squishy pandas to mini video game characters.

SPECTRAL CLOAKING

NOTHING TO SEE HERE.

"SPECTRAL" MEANS "GHOSTLY." So if you think "spectral cloaking" has something do with spooky magic, you're not too far off. It's a way to make things appear invisible! But instead of donning a magical invisibility cloak, scientists are using real tech to make the impossible possible. Ordinarily, light has to bounce off an object for us to see it. Spectral cloaking devices—which aren't clothes but rather boxy electronic gizmos with knobs and buttons—change light's frequency so that it goes through an object instead of bouncing off it. For now, a cloak works only when the thing being cloaked, the "target object," has its exact colors programmed into the cloaking device. Eventually, the tech could be used to disguise spy planes, cars, tanks, or even telecommunication waves so that secret conversations stay secret.

NERD ALERT: SCI-FI COOL

RESEARCHERS ARE DEVELOPING more ways to make things vanish. Plasmonic cloaking uses silicon nanowires coated in gold. When light hits these wires, the reflections from the gold and the silicon cancel each other out, making them invisible!

FUN FACT

Scientists in California, U.S.A., created a superthin invisibility "skin cloak." The only issue? It's microscopic!

FUN FACT

A Canadian company has been working on Quantum Stealth, a material that bends light to create invisibility. Right now, it's more like a flexible plastic sheet than a cloak, but it does make anyone behind it appear to vanish.

CYBERNETIC ENHANCEMENTS

A DIGITAL CHIP IMPLANT

THEY SOUND LIKE THE STUFF OF SCI-FI, but they exist. Quite simply, cybernetic enhancements are machines that become part of a person's body. Replacement hip sockets, implanted vertebral disks, and prosthetic limbs are all human-made pieces that become part of the body and improve or restore some of its abilities. In fact, some experts say that anyone who uses a device to search the internet is using it as an extension of their brain. To these experts, using a digital brain (like a smartphone) means that you're using a cybernetic enhancement. But if you're looking for superpowered humans more like the ones you see in movies, check out biohackers. They're people who install magnets, digital scanners, LEDs, and other tech in their bodies.

NERD ALERT: SCI-FI COOL

SOME OF THE LATEST WAYS to enhance your body are obvious, including implanted microchips that let you unlock your door by waving a hand. Some enhancements aren't obvious at all. They're microscopic. CRISPR-Cas9, an enzyme that cuts through DNA, might someday give people custom genes for things like giant muscles, crazy endurance, or disease resistance.

FUN FACT

"Wetware" used to be a joke word for human DNA, as opposed to software, which refers to computer programs. But today, it can also mean tech that's implanted in humans.

FUTURISTIC FOOD

YOUR BREAKFAST, LUNCH, AND DINNER COULD LOOK RADICALLY DIFFERENT IN THE NEXT FEW YEARS. SEE IF THIS IN-DEVELOPMENT EDIBLE TECH TICKLES YOUR TASTE BUDS.

OVALBUMIN EGGS

ON YOUR PLATE, they look exactly like scrambled eggs. But they don't come from chickens or even have shells to crack. Researchers make these faux eggs by putting chicken genes into microbes and letting the microbes create egg protein—aka ovalbumin.

APEEL

MASH UP SOME PRODUCE PEELS, seeds, and pulp; whip them into an edible coating and—voilà!—you've got Apeel, a protective coating for fruits and veggies that seals out oxygen and keeps in moisture, doubling the shelf life of your produce.

CELLPOD

IT'S LIKE A MINI GREENHOUSE, but instead of growing whole plants, CellPod produces clusters of plant cells that contain all the same nutrients as an entire plant. They aren't quite as tasty as real berries, but the vitamin-rich cells are ready fast—they can be harvested within a week. Open wide for mushy goodness!

SOLEIN

IT'S LIKE MAKING FOOD OUT OF THIN AIR. Seriously. Take some microbes and some air, zap them with electricity, and suddenly, you've got protein! It's a process Solar Foods calls gas fermentation. Gas fermentation pulls carbon dioxide (CO_2) and water from the air, and brings them together with microbes to make food (well, protein bits). Bonus: It reduces CO_2 in the atmosphere, so it makes you and the planet healthier.

SELF-DRIVING CARS

THERE'S NO DRIVER NECESSARY FOR THIS SELF-DRIVING CAR.

CLIMB INTO THE DRIVER'S SEAT, state your destination, sit back, and relax. Self-driving cars do the work for you—no license required. Also known as autonomous vehicles, these cars use LIDAR (that's pulses of laser light that reflect off surrounding objects) to calculate their direction, distance, and speed. They use cameras, sensors, and software to detect pedestrians, other vehicles, and obstacles on the road. And if that's not impressive enough, they can also predict what other moving objects are going to do based on their speed and direction. Of course, self-driving cars recognize and obey signs and traffic lights, too. Still think a human driver is safer? According to the National Highway Traffic Safety Administration, between 94 percent and 96 percent of all car crashes are caused by human error.

RADAR SYSTEMS HELP SELF-DRIVING CARS SENSE THE WORLD AROUND THEM.

NERD ALERT: SUPER SMARTS

WHY TRUST A SELF-DRIVING CAR? For starters, you can't see three football fields away, but self-driving cars can. They also don't get tired, distracted by the radio, or annoyed by your little sister—all things that can lead to accidents. But the programming has to be perfect, or else a self-driving car becomes just one more bad driver on the road.

FUN FACT

English company Aurrigo developed the first autonomous vehicle especially for people with health conditions and disabilities such as blindness. It will travel through airports, theme parks, campuses, and shopping centers.

HIGH-TECH HELPERS

IF YOU NEED A HAND, THIS ASSISTIVE TECH CAN DO THE JOB.

WCMX WHEELCHAIR

IN THE WORLD OF WHEELCHAIRS, WCMX—developed by Box Wheelchairs—is extreme. Stunt rider Aaron "Wheelz" Fotheringham set a bunch of world records in his chair, including the farthest ramp jump in a wheelchair—70 feet (21 m), which is twice as long as a telephone pole—and the first ever wheelchair backflip. WCMX's reinforced frame, racing shocks, and suspension make it tough enough to soar.

AAC

AUGMENTATIVE AND ALTERNATIVE COMMUNICATION (AAC, for short) devices give non-speaking people other ways to communicate. Some of the most popular devices include iPads and cell phones loaded with icons a person taps to say what they mean. For people with a limited range of motion, eye-gaze machines, like one known as Tobii, let users simply look at a letter, word, or icon to choose it.

EMBRACE SEIZURE BAND

IT LOOKS LIKE A SMART WATCH, but it has a lifesaving specialty: detecting seizures and calling for help. The watch picks up on electrical impulses and temperature changes through the skin, as well as unusual movement, to sense when its wearer is experiencing a seizure. Then it sends an alert to family or caregivers, who can rush to provide assistance.

LIFTWARE LEVEL

THE HANDLE ON THIS SILVERWARE robotically moves and adjusts for trembling hands, or to accommodate people who have a limited range of motion. Two internal motors bend and twist the spoon or fork so it stays level even when a person's hand doesn't.

GOT AN IDEA FOR SPECIAL TECH LIKE THIS? THE NERD OF NOTE ON THE NEXT PAGE MIGHT INSPIRE YOU!

NERD OF NOTE:
SADIE McCALLUM

Lots of us dream of going places. Well, New Hampshire, U.S.A., teen Sadie McCallum is making those dreams come true—by inventing things that help people with mobility issues get everywhere they want to go!

When she was nine years old, Sadie went to get a book from the library, but got an annoying surprise instead. "My town's library has these big steps," she says. "I was like, are you kidding me?" Sadie was born with cerebral palsy, which means her muscles don't work the way most kids' do. She uses a walker, but her walker couldn't climb the steps. So she had a problem.

Sadie mentioned it to teacher Deborah Lynch, who convinced her to try inventing a solution for the school's Invention Convention. "It didn't work at first," Sadie says. But she wouldn't give up. She tried making the wheels into a square shape ... then a triangle ... then loosened the bolts. At last, the Amazing Curb Climber, a walker with stair-climbing wheels, was born.

Excited, Sadie realized she might be able to solve more problems with inventions. For starters, she sometimes used a walker and sometimes a wheelchair, but it was hard to fit both in the car. So she built the Walker Wheeler, a walker that transforms into a wheelchair! Then she wanted to make it even better. So she came up with the WOW (Wonderful Omni Walker), a foldable, adjustable, walker-wheelchair hybrid that can also climb curbs.

And she wasn't done. Sadie knew she sometimes used the wrong muscles to walk, and that could lead to injuries. She designed the ATOM (Automatic Therapy Orthopedic Machine), a clip-on device that blinks and vibrates to prompt her to use the right muscles, and it won a patent award! Then last year, she and her sister, Claire, co-created LEGS (Liftable Exoskeleton Gait System), a therapeutic walking aid that helps users like her train their muscles to work more efficiently. Many of the components came from their local hardware store.

"If you want to invent, you don't need fancy materials," advises Sadie, who's now taking engineering classes. "You can make models out of cardboard. Especially if you have an idea that's going to help you or other people, that needs to be shown to the world."

Sadie tinkers with WOW wheels in her workshop.

"I've always asked, 'Yeah, but why?' and I think that's what makes me good at engineering ... I need to understand every component when I make a project."

SADIE MCCALLUM, inventor

ROBO HIGH

CLASS SUPERLATIVES

Most Likely to Go Far

PUTTING ONE FOOT IN FRONT OF THE OTHER ... and the other ... and the other, quadruped robot Xingzhe No. 1 became a track star by completing 1,405 laps around a track in China. With its jaunty walk, the bot sauntered along for 54 hours and 34 minutes on a single charge, racking up an incredible 83.28 miles (134.03 km). That's like walking the length of New York's Manhattan Island six times! The remote-controlled superathlete was created with the ultimate goal of walking into places that would be too dangerous for people, such as disaster sites. For now, Xingzhe No. 1 is practicing getting its steps in.

MONONOFU, **page 12**

Alpha 1S, **page 96**

Forpheus, **page 138**

Tradinno, **page 174**

ENIGMA MACHINE

YOUR ASSIGNMENT: Crack a code generated by a machine that can create billions of possible combinations. Sounds like a job for a superhacker, right? Back in World War II, the German military used an Enigma machine to encode messages that seemed unbreakable. Enigma employed substitution encryption, in which one letter stands for another, and the machine changed the code each time the same letter was pressed. There were more than 158,962,555,217,826,360,000 ways to set the machine! But the Germans always started their messages with a weather report and ended them with "Heil Hitler," and these clues helped decoders get cracking. Researchers, including mathematician and logician Alan Turing, finally broke the code after getting access to German codebooks and using them to design a Bombe machine—basically, a custom code breaker for Enigma.

NERD ALERT: VINTAGE VISIONARY

JUST AS TODAY'S PROGRAMMERS NETWORK multiple computers to solve a tough problem, the Bombe machine worked like 36 Enigmas wired together. By running them at once, it could crack a message in less than 20 minutes. Hackers call this as a "brute force attack," which means every possible combo is tried until one works.

THIS MACHINE MAY LOOK LIKE A TYPEWRITER, BUT IT'S ACTUALLY THE SUPER COMPLEX ENIGMA!

FUN FACT

Standing in front of the CIA grounds in Langley, Virginia, U.S.A., is a coded sculpture called Kryptos. It's so hard to crack that no one has been able to solve it, but some suspect it has to do with Enigma.

FUN FACT

One problem with Enigma was that it couldn't encode a letter as itself—you couldn't have *S* stand for *S*, for instance. That flaw was a big help to code breakers.

TOTALLY PAW-SOME!

THE REST OF YOUR FAMILY USES WI-FI, SO WHY NOT YOUR PETS? FOUR-LEGGED FRIENDS CAN GET HOOKED UP WITH THESE TECHIE CREATURE COMFORTS.

WHISTLE GO

IF YOU LOSE YOUR PHONE, you can track it. So why not your dog? Whistle Go is a GPS tracker that clips to a collar, so if Fluffy escapes, you'll get an email or text with her location. It also has a fitness tracker, so you can tell how much your pet is running, licking, scratching, or snoozing.

VARRAM PET FITNESS ROBOT

THIS BOT LOOKS LIKE WALL-E'S COUSIN, if Eve from *Wall-E* were part pet toy. Using an infrared sensor, this rolling bot can tease cats with a feather toy, handle rough play with pooches, and dispense treats, all while exercising your pet and keeping it (virtual) company.

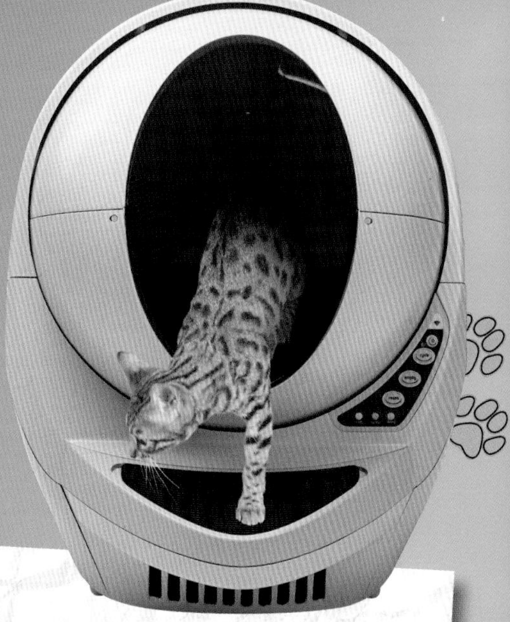

LITTER-ROBOT

WHO WANTS TO SCOOP CAT POOP? Nobody. That's why there's Litter-Robot, the robotic cat box that looks like a space module. Plus, it's Wi-Fi–enabled, so you'll know when it's ... occupied.

PETCUBE BITES

IF YOU HATE LEAVING YOUR PUP BEHIND when you go to school, check this out. Even when you're far away, you can toss treats and talk to your pooch through Petcube Bites' treat dispenser. Its app lets you stream video with night vision, zoom in, and fling treats up to six feet (1.8 m).

ELECTRONIC MOSQUITO REPELLENT

THE THERMACELL CREATES AN INVISIBLE, MOSQUITO-PROOF SHIELD.

SWAT ... SWAT ... MISSED IT! Mosquitoes can be annoying, and even be dangerous when the irritating bugs carry diseases such as Zika or West Nile virus. So using good repellent is no joke. But what if—instead of spraying yourself with stinky or toxic stuff—you could simply switch on a device? You can. Electronic mosquito repellents are stand-alone or carry-with-you devices that work by heating an odorless chemical repellent (derived from chrysanthemums) that they vaporize. The most popular electronic repellent, ThermaCell, produces an invisible cloud that repels skeeters for up to 15 feet (4.6 m) in all directions, no bug zappers necessary. And that's good news, because zappers kill beneficial bugs, too. This way, friendly moths aren't affected, while mosquitoes stay away.

NERD ALERT: Everyday Egghead

WHILE ELECTRONIC MOSQUITO REPELLENTS create a kind of invisible force field against some irritating insects, they can't stop bigger pests like rats. For that, there are sonic and ultrasonic soundmakers. The Guardian is one that also uses predator calls, a strobe light, and a thermal motion sensor to keep pests away.

FUN FACT

You can buy clothing with built-in mosquito repellent.

FUN FACT

There are lots of anti-mosquito apps that claim to keep away bugs by using sound, but there's no evidence that any of them work.

WHAT'S YOUR DREAM TECH?

WHICH TECH WOULD MAKE YOUR WILDEST DREAMS COME TRUE? LET'S FIND OUT!

START HERE

ARE YOU AN OUTDOORS ADVENTURER OR A HOMEBODY?

I like to hang out at home—with my tech.

I want to see the world.

WOULD YOU RATHER HAVE A COMPUTER THAT ANSWERS ALL YOUR QUESTIONS, OR ONE THAT MAKES YOUR FAVORITE "IMAGINARY CREATURES APPEAR"?

I want to see imaginary creatures!

I'm full of questions, and would love answers.

HOW DO YOU LIKE TO TRAVEL?

Road trips are awesome.

SELF-DRIVING CARS
Your dream tech is a self-driving car!

In a plane!

HOW DO YOU FEEL ABOUT SUPERHEROES?

AI
Your dream tech is artificial intelligence!
I want to be one!
I'd rather fly a super techie plane.
JET PACK
Your dream tech is a jet pack!
AUGMENTED REALITY
Your dream tech is augmented reality!
VTOL
Your dream tech is a vertical take-off and landing vehicle (VTOL)!

FREEWHEELING FOOD

BELLY RUMBLING? CLICK A BUTTON, AND ONE OF THESE ROBOTS ROLLS TO THE RESCUE.

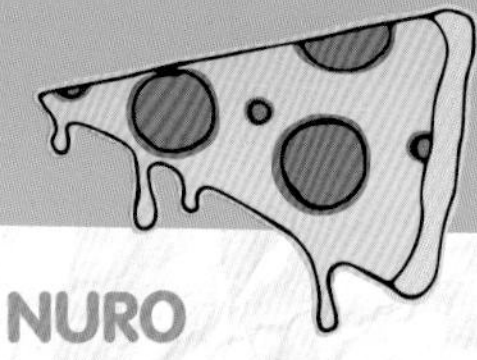

NURO

IT LOOKS LIKE A MINI MINIVAN, but the only things inside are food. To driverlessly bring you meals or groceries, Nuro's vehicle uses sensors, 12 cameras, radar, and a 3D map to "see" road features such as curbs, lanes, traffic signs, and speed bumps.

STARSHIP

THIS BLACK-AND-WHITE VEHICLE rolls in like a mini Stormtrooper to bring you pizza, drinks, and doughnuts within a four-mile (6-km) radius. Wondering where your grub is? Check the app to track your bot, which is being "employed" by popular restaurant delivery companies. And don't worry about theft. The bot has 360-degree cameras, and it only opens with your smartphone.

SNACKBOT

SPORTING COLORFUL WHEELS and a jaunty flag, this bot cruises through college campuses carrying snacks and beverages. With its camera and headlights to help it navigate through darkness or rain and all-wheel drive to climb steep hills, it's on a mission to deliver your munchies.

PAC-MAN

BACK IN 1980, most video games were about shooting aliens. Toru Iwatani, a Japanese video game creator, set out to make a family-friendly arcade game that was all about fun, with no violence. Inspired by cute cartoons, he designed a maze game that would be about something everybody likes—eating. With only a joystick, it would be easy enough for young kids to try but challenging enough for grown-ups to keep playing. And unlike other games of the time, this one would have a main character. Iwatani got the idea for his character when he took a slice from a pizza and realized the remaining pie looked like a munching monster. The game, Pac-Man, was a hit worldwide, gobbling up one billion dollars in America alone within the first 15 months. Pac-Man's golden face showed up on everything from T-shirts to lunchboxes, telephones, and even a hit record, and he munched his way to becoming the most successful arcade game of all time.

NERD ALERT: VINTAGE VISIONARY

PAC-MAN'S BLOCKY GRAPHICS MIGHT SEEM OLD-FASHIONED NOW, but they were using the best data processor available at the time, called 8-bit. One bit is the smallest amount of info a computer can store and access at a time. That means computers could process just eight bits of data—about enough to generate a single alphabet letter or number—in one go. Eight-bit paved the way for today's computer processors, which are about 72 quadrillion times faster!

AN EXAMPLE OF 8-BIT GRAPHICS

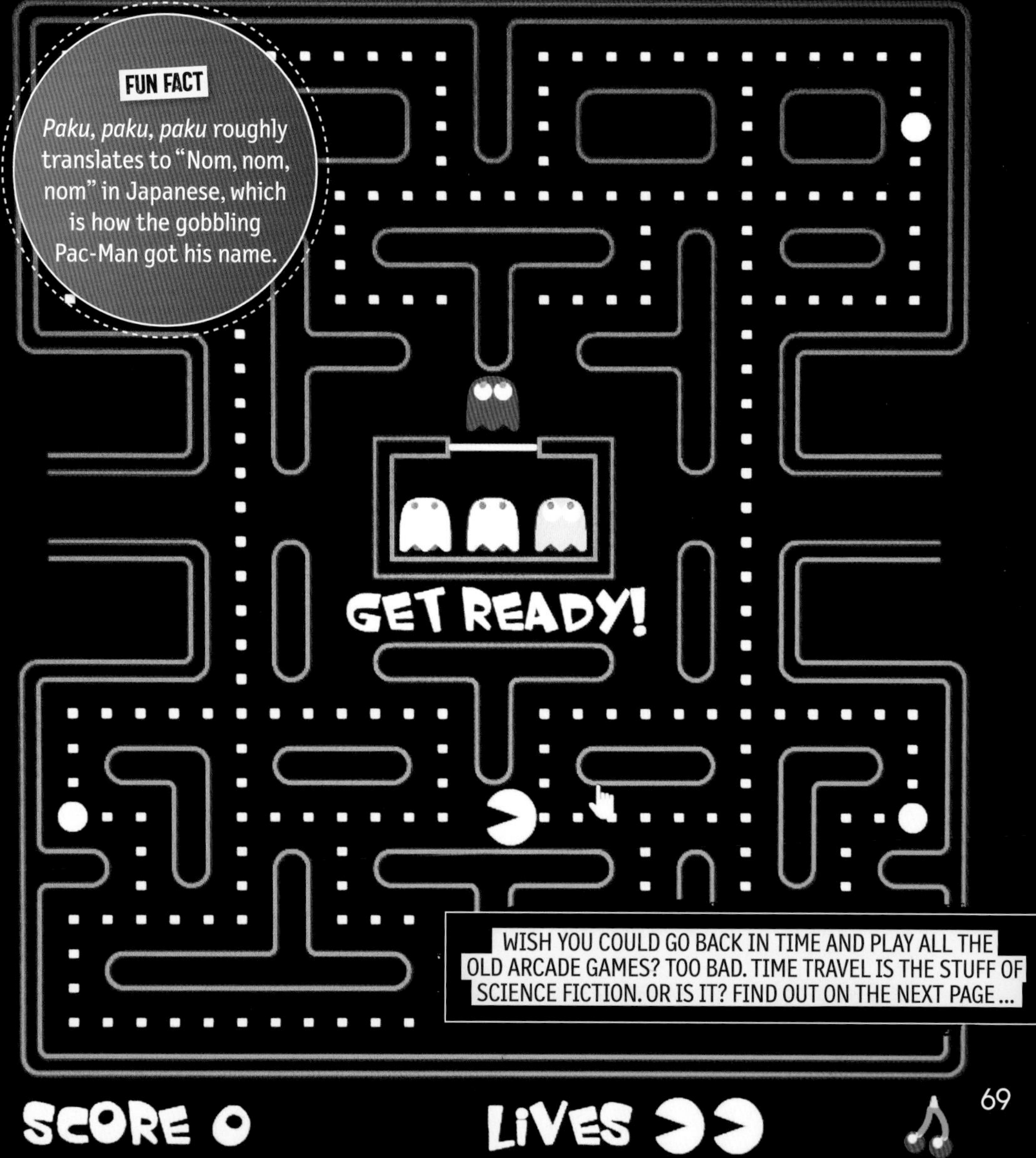

FUN FACT

Paku, paku, paku roughly translates to "Nom, nom, nom" in Japanese, which is how the gobbling Pac-Man got his name.

WISH YOU COULD GO BACK IN TIME AND PLAY ALL THE OLD ARCADE GAMES? TOO BAD. TIME TRAVEL IS THE STUFF OF SCIENCE FICTION. OR IS IT? FIND OUT ON THE NEXT PAGE ...

FAQ COULD YOU EVER TRAVEL BACK IN TIME?

Science fiction makes it look easy: All you have to do to zip through time is find the right tool. But could backward time travel happen in real life? The answer, theoretically at least, is yes. You'd first need a "quantum wormhole," which could possibly be made by connecting a black hole to something with an equally large negative mass. (Researchers have already created negative mass atoms by supercooling the metallic element rubidium.) If scientists could generate enough negative mass, they could make a wormhole and bend space-time.

FUN FACT

Time travel is all over fiction—Doctor Who's TARDIS, Hermione Granger's time-turner, and the tesseracts from *A Wrinkle in Time* and *The Avengers* are just a few examples.

Thing is, no one knows where or when you'd wind up if you went through. Time and space are connected—as Albert Einstein's theory of relativity describes it, everyone experiences where and when things happen depending on how fast they're moving—and things really get warped when you move close to the speed of light. So once you traveled through the wormhole, you might find yourself far across the galaxy or even in another universe. And because of time's distortion through space-time travel, when you tried to return home, you might discover your world has aged decades ... or more. Finally, there's the little issue that a black hole's gravity squashes anything that enters it. Until we solve these problems, you won't be needing your space-time passport.

FUN FACT
Earth's clocks tick slower than satellite clocks. That's because time is slowed slightly by Earth's gravity.
FUN FACT
The first author to use the term "time machine" was H.G. Wells in 1895.

DEEPSEA CHALLENGER

JAMES CAMERON OPERATES HIS SUBMERSIBLE FROM THE PILOT'S CHAMBER.

IF YOU DREAM OF VENTURING INTO AN UNMAPPED WORLD, space isn't your only option. Some of Earth's greatest mysteries lie under the sea. So what's down there? In 2012, National Geographic Explorer James Cameron went to find out. He took a vertical torpedo submersible almost seven miles (11 km) down to Challenger Deep, the deepest known spot in the ocean. The sub, named *DEEPSEA CHALLENGER*, had to handle crushing underwater pressure and near-freezing water—all with zero sunlight. It was equipped with a robotic claw, a sediment sampler, a "slurp gun" to suck up sea creatures and observe them, and 3D video cameras. Cameron found giant, shrimplike crustaceans, massive sea cucumbers, and "microbial mats"—layers of microorganisms that live off chemicals emitted by rocks. These microorganisms may teach scientists how all life began.

NERD ALERT: SCI-FI COOL

TO ALLOW *DEEPSEA CHALLENGER* to handle the immense pressure of being nearly seven miles (11 km) underwater—15,000 pounds per square inch (1,055 kg/sq cm)—70 percent of its chassis was made of foam that could compress without crushing. But that meant the sub also needed a pair of steel weights totaling 1,000 pounds (454 kg) to keep it underwater. When it was time for the sub to surface, the weights were dropped by electromagnets.

FUN FACT

The capsule Cameron sat in was spherical because that shape is best equipped to resist pressure. Cameron's capsule wasn't as equipped for comfort, though: He had no room to stretch his legs or arms while inside.

FUN FACT

Though he was alone in his submersible, Cameron wasn't lonely. He could communicate with the surface by voice or text message.

SPY CAMS

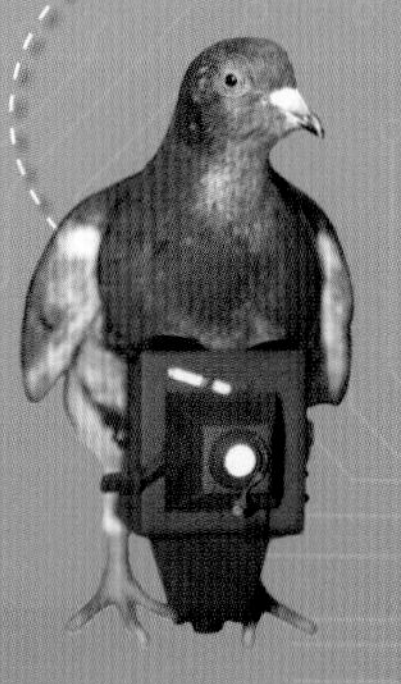

THAT'S NO AVERAGE PEN IN THIS POCKET—IT'S A SPY CAM!

WHAT DO WATER BOTTLES, lipsticks, alarm clocks, and jewelry have in common? They're all places you can hide a miniature spy camera. These days, spy cams use Wi-Fi and run on batteries, so there are no wires to hide. Some are also equipped with motion detectors that make the cameras move when someone passes them. If that isn't sneaky enough, some project sound waves too high for humans to hear and are triggered when someone walks through the waves. Others get tripped when they sense body heat. And still others send out lasers and wait for someone to cross them. Today, the police and the FBI can catch crimes being carried out by using cameras hidden in trees, rocks, tombstones, and even vacuum cleaners!

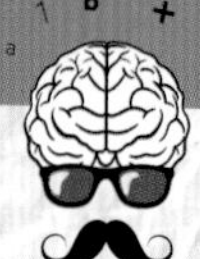

NERD ALERT: SUPER SMARTS

WHILE COUNTERINTELLIGENCE EXPERTS now have hidden camera–detecting apps to help them spot spy cams in action, they also have clever, low-tech ways to uncover undercover spies. They listen for soft whirring—the sound of a hidden camera moving. They also look for holes in walls. If they shine a light off any of these holes and see the glint of a lens ... caught!

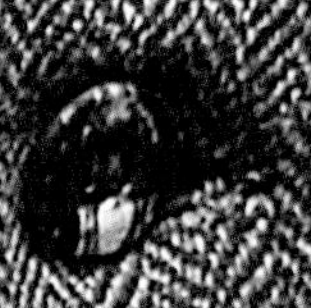

FUN FACT

Since as early as 1907, pigeons have been used to carry spy cameras and get footage from above.

FUN FACT

Around 1960, Soviet secret police wore cameras hidden in their coat buttons. They could press a button inside a pocket to snap a picture.

NERDSVILLE CENTRAL: INTERNATIONAL SPY MUSEUM

AN UNDERCOVER ATTRACTION

IF YOU'VE EVER DREAMED of having an unusually talkative spy for a friend—say, one who'd give you a peek at undercover spycraft and let you test out the tech—this Washington, D.C., museum might help your dream come true. After all, founding executive director Peter Earnest spent 35 years with the CIA before setting up the museum to reveal all the real-life (secret!) stuff he could legally tell you about. Feast your eyes on lock-picking kits, declassified CIA tech like lipstick pistols, an Enigma machine (read more about it on page 58!), and other super-sly, high-tech spy gadgets. You can even see a car from the most famous fictional spy of all, James Bond.

1 Prepare for your mission. When you arrive, you can launch a mission, during which your performance will be tracked by a radio frequency-enabled badge you wear. You'll even get a code name.

2 Think gadgets. You'll get to digitally design your ideal spy gadget at an interactive exhibit, so consider what kinds of gizmos would be your dream spy machines.

TIPS TO NERD OUT LIKE A SPY:

3 Be strong. At the "Hang Time" exhibit, you'll find out how long you can hang from a helicopter bar, with "wind" blowing around you, like a real superspy!

4 Pay attention to details. At the end of your mission, you'll get a memory quiz to uncover your top spy skills.

ISS

WHAT'S NERDIER THAN A RESEARCH LAB? A research lab ... in outer space! And, on certain nights, you can see it just by looking up. That bright light in the sky is the International Space Station (ISS), essentially a science building about the size of a five-bedroom house, orbiting 240 miles (386 km) above Earth. Inside, crews of astronauts run experiments to learn more about how things act in space. They've tested 3D printers, high-definition televisions, airborne particles, wearable sensors, plants, and much more, with the goal of using what they learn to help people travel deeper into space. The station itself is a technological marvel: Its special features include solar energy–gathering "wings" called solar arrays, and robotic arms that help with repairs and pull things from space into the ISS. (Those things could be anything from science experiments to toilet paper for the astronauts.)

CANADARM2 IS A ROBOTIC ARM THAT HELPS WITH SPACE STATION MAINTENANCE.

FUN FACT

To spot the ISS, look for the brightest object in the sky with no flashing lights. It circles Earth every 90 minutes.

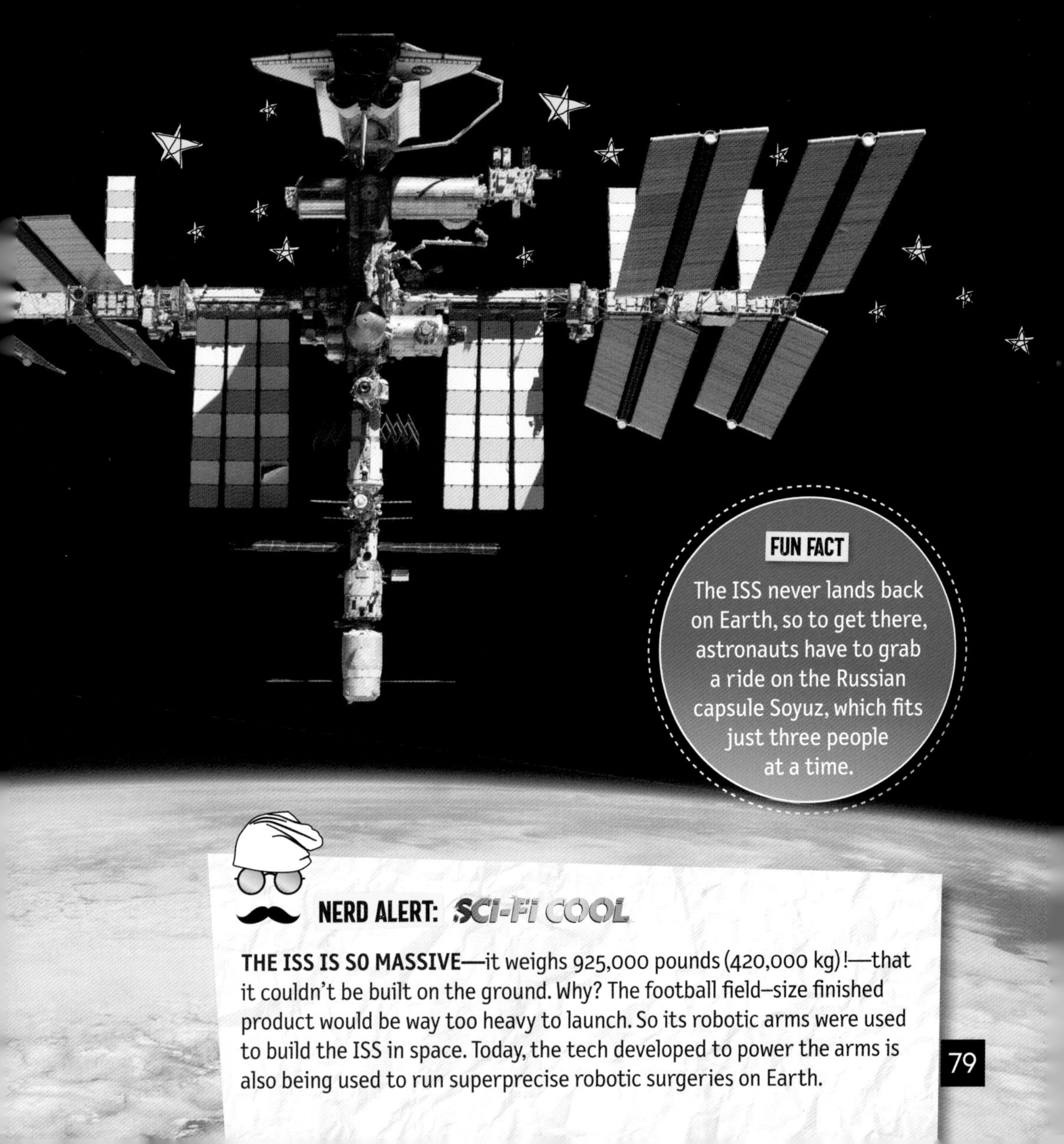

FUN FACT

The ISS never lands back on Earth, so to get there, astronauts have to grab a ride on the Russian capsule Soyuz, which fits just three people at a time.

NERD ALERT: SCI-FI COOL

THE ISS IS SO MASSIVE—it weighs 925,000 pounds (420,000 kg)!—that it couldn't be built on the ground. Why? The football field–size finished product would be way too heavy to launch. So its robotic arms were used to build the ISS in space. Today, the tech developed to power the arms is also being used to run superprecise robotic surgeries on Earth.

PERSON-LESS PLANES

THESE UNMANNED VEHICLES CAN ENTERTAIN YOU, HELP THE ENVIRONMENT, OR EVEN SAVE YOUR LIFE. MEET SOME EXCELLENT DRONES.

VERGE AERO

THESE BRIGHT LIGHTS aim to one day replace fireworks by offering a safer way to do amazing light shows. The drones can be programmed to make pictures in the sky, and have already made their music video debut with Australian trio PNAU.

TREE-PLANTING DRONE

ALMOST HALF OF THE WORLD'S MANGROVE FORESTS have been slashed to make room for farms, factories, harbors, and houses. That's bad news because mangroves help stabilize coastlines, dampen storms' impact on land, and provide homes to animals. In Myanmar, drones are swooping in and shooting seeds that, eventually, sprout into trees. Ten of these do-gooder drones could plant up to 400,000 trees in a day!

SAILDRONE

AHOY! Running on wind and solar energy, the Saildrone collects ocean data to help monitor Earth's health. One recently completed the first autonomous trip around Antarctica, surviving 49-foot (15-m) waves and dodging icebergs, to gather info like water temperature and acidity, air pressure, wave heights, and more.

LITTLE RIPPER LIFESAVER

THIS SOLO FLIER comes to the rescue by delivering lifesaving equipment such as rafts, thermal blankets, mobile defibrillators, first aid kits, and food to people in need. It also has artificial intelligence and a camera, allowing it to, among other abilities, alert swimmers to nearby sharks.

EVTOL

FLYING TAXIS COULD ONE DAY ZOOM OVER TRAFFIC JAMS.

IT'S A HELICOPTER! It's a plane! It's a ... really weird hybrid helicopter plane? The eVTOL—short for "electric vertical takeoff and landing" vehicle—is an electric plane that rises straight up like a helicopter or drone, and also hovers. Some use as many as 12 small propellers along their wings—technology called distributed electric propulsion (DEP), which basically means spread-out electric motors. Others, like the Pipistrel 801, use fans for lift. And some eVTOLs can also drive. The ASKA, which means "flying bird" in Japanese, is an in-the-works eVTOL car. It can cruise along highways and also lift off into the sky, thanks to electric fans all over the hood and trunk, and wings that fold out, almost like origami.

NERD ALERT: SCI-FI COOL

ENGINEERS EXPECT THAT THE TECH BEHIND EVTOLS will change how planes are designed in the future. Their wings won't have to tilt backward like the wings of jet planes. In fact, some won't have wings at all, but instead will look more like drones. And because electric planes are quieter than jets, neighborhoods around airports could end up sounding a lot different.

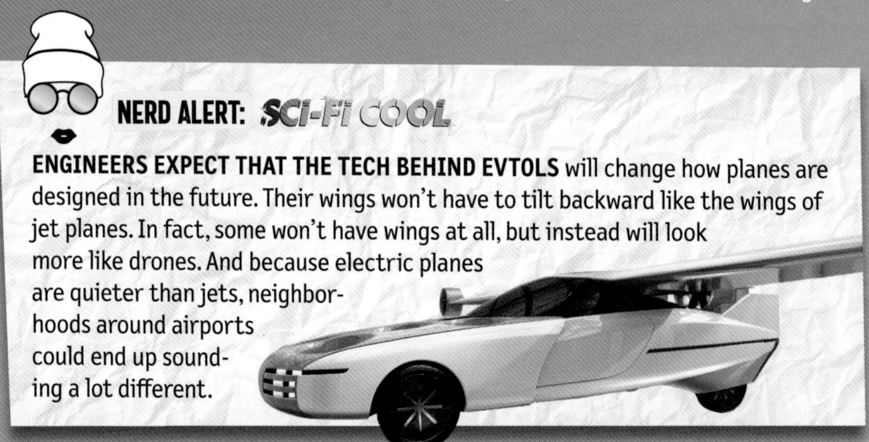

FUN FACT

eVTOL makers hope you'll soon be able to reserve a flight through an app and catch your "air taxi" on top of a local building.

HYDROGEN FUEL CELLS

FUEL CELL

A TOYOTA MIRAI ENGINE

O_2 H_2 H_2 H_2O e^- H^+ FUEL CELL

A HYDROGEN FUEL CELL

WHAT IF THE WORLD COULD SAVE much needed energy by using the most plentiful element on Earth? Maybe it can, say creators of hydrogen fuel cells. The cells are battery-like devices that convert hydrogen-rich fuel into electricity through a chemical process that's almost silent, and its only by-products—the stuff that's left over—are heat and water. They work by sending hydrogen through the positively charged end of the cell (the anode), and oxygen through the negatively charged end (the cathode). The anode splits hydrogen molecules into electrons and protons. Then the electrons are pushed through a circuit to generate a current. Hydrogen fuel cells can be used to run electricity in buildings, give juice to your favorite devices, power up special cars—like Toyota's Mirai—and more.

MIRA

NERD ALERT: DO-GOOD GEAR

FUEL CELL VEHICLES could reduce CO_2 emissions by 90 percent. So why aren't they more common? Mostly, the world isn't prepared for them yet. We don't have enough dedicated pipelines to transfer hydrogen or enough ways to store it. Plus, escaped hydrogen could harm the atmosphere, so when pipelines and storage are created, they have to be really good.

FUN FACT

One U.K. company is reusing plastic trash by heating it until it produces hydrogen for fuel cells.

IF SWITCHING UP EVERYDAY TECH TURNS ON YOUR CURIOSITY, CHECK OUT THE NERD OF NOTE ON THE NEXT PAGE.

NERD OF NOTE:
MARTIN MOLIN

Plunk. Plink. Plunk. If you've ever played with a marble coaster, you know that satisfying sound. So does Swedish musician and maker Martin Molin. And he decided to take his marble coaster as far as it could go ... by building it into a stadium-ready rock music instrument!

Growing up, Martin loved building with Lego bricks. "I am still totally fascinated by complex systems and constructions," he says. "And cogs. I love cogs!" When he was 13, he saw a video of rocker Jimi Hendrix playing live. "It was like he had created a totally new sound library," says Martin. "It opened my eyes to how technology could be used to make new sounds."

He went on to study music in college, and in 2011, he co-founded the band Wintergatan. Three years later, Martin was working on new music when he started to feel stuck. He decided to break up the work by building a new instrument.

After 16 months, Martin's 6.5-foot (2-m)-tall musical Marble Machine was done. Martin turns a crank to activate the gears. These release 2,000 marbles, which fall on instruments that include a bass drum, a vibraphone, and a snare drum whose sound comes from dry rice.

Martin's latest creation is Marble Machine X, a version that can be disassembled and put back together so it can go on tour. "As long as you have a dream, you can succeed," says Martin. "That's what I think."

"I have a strong need to connect to people through the things they create and the things I create ... We're all a small gear in a big machine, in a way."

MARTIN MOLIN, musician and maker

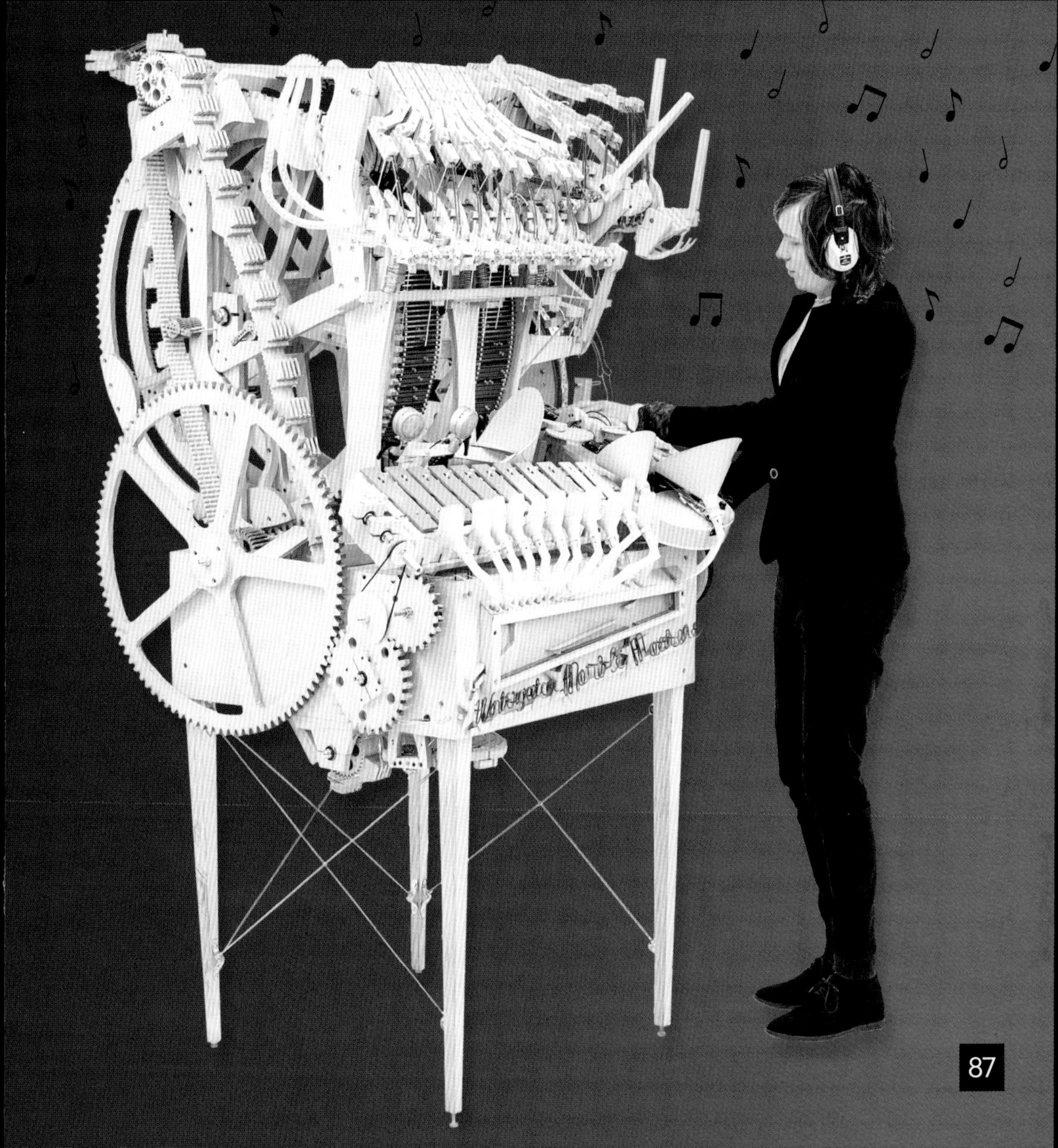

CLONING

FUN FACT

A cloned cat named CC was genetically identical to her mom, but her coat pattern was different. Coat coloring comes from pigment cells in the skin, which move around during fetal development.

CLONES HAPPEN IN NATURE ALL THE TIME—consider identical twins, which are natural clones of one another. But "artificial" clones of whole animals are made by scientists. Researchers start by removing the nucleus from a female mammal's unfertilized egg. Then, in a totally sci-fi move, they take one of the to-be-cloned animal's cells and either inject it into the egg, or fuse it with the egg using electricity. Cell division begins, an embryo is created, and this embryo is implanted in a surrogate mother. It grows until ... ta-da! A clone of the original animal is born! Some people, such as farmers, would like to clone animals with valuable traits like high-protein milk. Conservationists might want to clone endangered animals to help preserve species. But not everyone agrees with cloning. Clones often have shortened lives, or are born with health issues. Plus, reproducing identical animals could cause a species to become too genetically similar to survive.

NERD ALERT: SCI-FI COOL

IF SCIENTISTS EVER FIND DINOSAUR DNA, cloning technology might be what's used to bring the giant lizards back. But should we bring them back, even if we could? One problem is that the food dinosaurs need to eat may not exist today. Also, where would you put such a big creature? And, almost as important, who'd scoop the dino-size poop?

FUN FACT

The world's first clone of an adult mammal was created in Scotland in 1996. After 277 tries, Dolly the sheep was born.

CHAT

YOU'VE PROBABLY WONDERED what animals would say if they could speak. Well, these inventors wondered the same thing! An animal communication team in the United States at the Georgia Institute of Technology in Atlanta, chose to focus their efforts on dolphins. The team invented CHAT—cetacean hearing and telemetry—a toaster-size box that deciphers dolphins' whistles. It's already translated one dolphin's whistles to mean a type of seaweed! Here's the catch though: CHAT can't yet pick up and decode dolphins' natural sounds. The translated seaweed whistle actually came from a sound researchers taught the animal to mimic. And it's even possible that the dolphin didn't understand what it was saying, but only repeated what it had learned. That said, researchers hope to ultimately use pattern recognition software to decode dolphins' own language.

DOLPHIN RESEARCHERS IN THE BAHAMAS SPORT THEIR CHATS.

NERD ALERT: SUPER SMARTS

WHILE IT MIGHT SEEM STRANGE that dolphins copy made-up sounds, it's actually in their nature. They've been known to mimic other dolphins' whistles in the wild, and to associate these whistles with objects or other creatures. Each dolphin has a "signature" whistle it uses to identify itself, kind of like its name, and dolphins use these whistles to call to their friends!

FUN FACT

The first animal "words" ever decoded were spoken by prairie dogs. They make specific sounds to identify different humans, and even what colors they're wearing!

FUN FACT

The FIDO Project (Facilitating Interactions for Dogs with Occupations) is teaching dogs to use sensor vests to speak. By turning different directions or pulling a cord, canines can make the vest say "I heard the doorbell," "I heard an alarm," or "My owner needs your help!"

iMAGINE THIS

PET CHAT

IT'S THE YEAR 2040, AND ANIMAL COMMUNICATION APPS HAVE BEEN PERFECTED. NOW, LET'S HEAR WHAT YOUR PETS ARE REALLY SAYING ...

input:
TWEET!
translation:
"Look out!"

input:
Twee-twee-twee-twee-tweet!
translation:
"Stay away from my nest!"

input:
Tweet tweet tweet!
translation:
"Turns out, it's the delivery guy. Who ordered anchovies?"

DOGS BARK HIGH AND FAST TO TELL YOU SOMETHING'S UP, AND LOWER TO TELL YOU THEY'RE SERIOUS. AND THAT "HA ... HA ..." PANTING SOUND DOGS MAKE WHILE THEY'RE PLAYING IS A LAUGH!

BIRDS HAVE SHARP, LOUD CALLS TO SIGNAL DANGER; A FAST TRILL TO PROTECT THEIR NESTS; AND LOUD, URGENT CALLS TO LOOK FOR OTHER BIRDS.

input:
Meow!
translation:
"Servant!"
input:
Meow. Meow!
translation:
"You are under my power."
input:
Meow. Meow. Meow!
translation:
"Whenever you hear my voice, you will dispense food. NOW!"
CATS TRAIN THEIR OWNERS. IF YOU KNOW WHEN YOUR CAT WANTS FOOD AND WHEN YOUR CAT WANTS OUT JUST BY ITS SOUND, YOU'VE BEEN TRAINED!
input: Woofwoofwoof!
translation: "Look out!"
input: Woof ... woof ... woof.
translation: "I'm serious."
input: Arf-arf-arf-arf! Ha, ha, ha!
translation: "I made the cat's hair stand up! Ha, ha, ha!"

SNOTBOT

SOME ROBOTS BUILD ROCKETS, some study climate change, and some ... collect whale snot. Seriously. The SnotBot hovers above surfacing whales and captures the mucus they blow out. The samples SnotBot collects show researchers whether the whales have any toxins or bacteria in them, and they're used to measure hormone levels that can reveal whether the whales are stressed. This info helps researchers track endangered whales' health and their reproductive cycles, and judge how well conservation efforts are going. Before SnotBot, researchers had to use noisy motorboats to chase down whales, then shoot them with sampling darts. In contrast, SnotBots fly far above the whales to take their samples, never touching the whales at all. Say "Achoo!"

FUN FACT

Right whales, which SnotBot studies, swim at about six miles an hour (10 km/h). That's about as fast as Olympic swimmer Michael Phelps' top speed!

NERD ALERT: DO-GOOD GEAR

SNOTBOT ALSO TAKES PHOTOS AND VIDEO to record every whale's unique fluke, or tail fin. Software with artificial intelligence "looks" at every pic, compares it to a whale research database, and identifies an animal if it's been photographed before. The SnotBot's camera also has night vision, so it can help catch poachers.

SNOTBOT FLYING
OVER A WHALE
FUN FACT
The stuff that comes
out of a whale's blow-
hole isn't ocean water.
It's mostly exhaled
breath—and snot,
of course!
FUN FACT
The New England
Aquarium in Boston,
Massachusetts, U.S.A.,
is working on a project
called Counting Whales from
Space that uses satellites
to help count and
map whales.

ROBO HIGH
CLASS SUPERLATIVES

Most Likely to Star in a Music Video

AT JUST A FEW INCHES TALLER THAN A FASHION DOLL, Alpha 1S really shines when it's with its buddies—1,371 other Alpha 1S's! The mega dance team strutted its stuff in Rome, Italy, during a music festival, and has also appeared internationally with live performers. From toe-tapping to backbends and break dancing moves, the robots use identical programming to stay in step. Alpha 1S is actually a singing and dancing toy created to help kids learn how to program. When it's not dancing, it does yoga and martial arts. Because hey, when you've got it, flaunt it.

MONONOFU, **page 12**

Xingzhe No. 1, **page 56**

Forpheus, **page 138**

Tradinno, **page 174**

MICROWAVE

THEY MIGHT SEEM LIKE BORING, everyday items, but—trust us—microwaves are modern marvels. They use strong radio waves called, well, microwaves to cook your food super fast. The waves penetrate straight into the food's particles rather than gradually heating the food from the outside the way a traditional oven does.

The secret to the microwave oven's speed is its magnetron. Inside the oven, this magnetron sends electrons through electric and magnetic fields, creating microwaves. These waves make the food molecules vibrate, and the faster they go, the hotter they get. Microwaves make water molecules spin especially wildly, which is why the wet insides of your potpie can be red-hot while the dry crust stays cool. Also, microwaves can't reach deeper than about an inch (2.5 cm) into food, after which the hot outer layer, rather than the actual microwaves, warms the inside.

THE RADARANGE

FUN FACT

The first microwave, called the Radarange, was five feet (1.5 m) high and looked like a regular oven.

NERD ALERT: Everyday Egghead

DOES YOUR MICROWAVED FOOD COME OUT WITH HOT AND COLD SPOTS? This could be because of "interference," which is what happens when microwaves bump into each other. Put your plate on the turntable's edge, not in the middle, and it'll cook more evenly.

FUN FACT

In the 1940s, engineer Percy Spencer was working with a magnetron at Raytheon Manufacturing Company when a peanut cluster bar in his pocket melted. He realized what the magnetron could do and got the idea for the microwave.

OK, MICROWAVES MIGHT BE MORE HIGH-TECH THAN WE THOUGHT. BUT ARE THEY SAFE? WE'VE GOT ANSWERS ON THE NEXT PAGE.

FAQ IS IT OK TO STAND IN FRONT OF A MICROWAVE?

Grown-ups might tell you not to stand there, but are they right? First, the bad news: Your microwave oven may leak radiation. But it's a maximum of five thousandths of one watt. For perspective, cell phones can safely emit up to 1.6 watts. And you'll only be affected by your oven's microwaves if your head is two inches (5 cm) away from the oven. The metal mesh lining your microwave stops waves from getting out, and the door latch has an "interlock" that shuts off the microwave as soon as you open it. Long story short, standing in front of your microwave while you're waiting for your popcorn to pop isn't going to hurt you.

If you really want to be careful around microwaves, you'd do better to watch out when you reach for the drink you just heated up. That's because water inside a microwave often doesn't bubble. Instead, it can "superheat," which means heat past its boiling point without bubbling first. Jostling the cup while you take it out can (in rare cases) cause bubbles to form so quickly that they blow up at you. So, if you really want to stay microwave safe, leave a little space between you and your beverage.

FUN FACT

One of the first experiments the microwave's inventor did was to put an egg in its shell under a magnetron ... where it exploded.

FUN FACT

Microwaves are also used to broadcast television.

FUN FACT

Microwaves bounce off mirrors. So if a supervillain released microwaves at you, you could deflect them with a mirrored shield.

CODE

COMPUTERS CAN SEEM EERILY SMART. They answer your questions and guess your favorite songs. But the truth is that computer "brains" are programmed by humans using code. Code is what gives a computer its instructions. You can write code in hundreds of programming languages, and programmers pick the ones that work best for what they want to do. JavaScript, for example, can give a website cool 3D effects, while Python is made to be simple enough that beginning programmers can use it to build games. In most cases, the computer turns the code you write into machine code, which means it translates it into a series of numbers—typically 1s and 0s—that stand for actions. Then the computer "executes" the code, doing exactly what it's been told to do. It can't guess anything, or read between the lines like humans can. So if your code tells a computer to make a milkshake, and you forget to include "stop"... well, you might have a big mess on your hands.

FUN FACT

In the mid-1800s, Ada Lovelace wrote the world's first computer code. It was for the Analytical Engine, a mathematical computing machine.

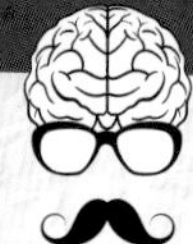

NERD ALERT: SUPER SMARTS

IN 1996, a computer called Deep Blue was coded to become a chess champion, and it beat chess legend Garry Kasparov. Deep Blue's coding was straightforward—it used all possible combinations of moves until it won. But today's computers are programmed to learn as they play, so even their programmers don't know what they'll do next.

FUN FACT

People who create malware codes that hurt computers often make these codes public after using them. They aren't being generous. They're giving their codes to more criminals so police can't guess who used them first.

NERDSVILLE CENTRAL: NATIONAL MUSEUM OF COMPUTING

A PC LOVER'S PARADISE

DID YOU KNOW THAT COMPUTERS WERE ONCE SO BIG, they each filled a room? It's one thing to hear about it, and it's another to actually see it in person. Fans of bits and bytes can get their hands on computer history at the National Museum of Computing in Milton Keynes, Buckinghamshire, England, which holds the world's largest collection of historical computers. Yes, working—you can see these massive machines humming away and also heating up the place. (Computers that big generate a lot of warmth!) There's Colossus, the world's first electronic computer, which helped decipher encrypted messages between Hitler and his generals. There's an air traffic control computer that lets you simulate takeoffs and landings. And there's a 1970s PC gallery, where you can try out the kinds of computers your grandparents used. Be sure to look near the PC gallery for arcade games you can play, including Space Invaders and Crazy Taxi!

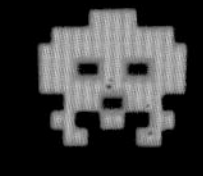

TIPS TO NERD OUT AT THE NATIONAL MUSEUM OF COMPUTING:

1 Check out the library. By appointment only, super computer nerds can dive deep into the museum's archive of books, manuals, and more.

2 Sign up for a "relaxed opening." These special sessions include hands-on activities and a chill-out zone for anyone prone to sensory overloads.

3 Come for a byte. On school holidays, the museum offers STEM Bytes festivals, where you can delve into Legos, Minecraft, virtual reality, and more.

4 Keep your little siblings busy. Your parents can ask for a family pack that includes a toy, book, and special map that'll occupy little kids while you're checking out the mainframes.

WI-FI

YOU PROBABLY USE IT EVERY DAY: Wi-Fi starts with an antenna sending data-filled, high-frequency radio waves. These radio waves beam to your router, which is a kind of data decoder. The router transmits freshly decoded data to your computer, which receives the information through a special Wi-Fi chip. The chip turns the data into web pages, email, or the latest meme. And all you have to do is log on.

But there's more: The Wi-Fi chip in your computer doesn't just receive info. It can beam it out, too. When you press "send" on a message or play a game online, your Wi-Fi chip turns your data back into a radio frequency and sends the frequency to the router, and off it goes to the antenna, to be received by another computer.

NERD ALERT: Everyday Egghead

WI-FI MAY SEEM ALMOST MUNDANE TODAY, but it's a modern marvel with an equally marvelous past. During World War II, Hollywood actress Hedy Lamarr attended meetings with her husband, an Austrian weapons manufacturer. Their marriage soured, but he didn't imagine that Lamarr, who was also a supernerd, was listening to their plans to jam American military signals. She realized that if radio frequencies keep changing, they won't be detected by enemies. Years later, the tech she invented, called frequency hopping, would be used in Wi-Fi.

HEDY LAMARR

FUN FACT

Wi-Fi is also called 802.11 networking. Why 802? Some say it's because the Institute of Electrical and Electronics Engineers (IEEE) created it in February of 1980 (February is the second month). The .11 might stand for a committee that worked on it.

DO YOUR DEVICES EVER GET TOO HOT TO HANDLE? THE NEXT PAGE COULD HELP COOL THINGS OFF.

FAQ WHY DO MY DEVICES GET HOT?

You're on an epic video chat with your BFF when you notice that your phone is getting pretty warm. Is it going to fry you? Or fry itself? While an old battery can heat up, usually it's the phone's processor and screen that get toasty. High-intensity apps like games heat up your phone's central processing cores and its graphics processing unit—and even your screen as it emits light. Is your brightness all the way up, or do you have lots of apps running? That can do it. Hot days can also overheat your phone just like they overheat you. None of this is likely to hurt your phone, but especially in the sun, it can trigger it to stop running until it cools. So, what should you do? Take off its case, and blow on it or fan it. Close apps, turn off Wi-Fi, and restart. And make sure your device charges on a smooth, cool surface, not on a pillow or a mattress that traps heat. Still, if your phone is getting so hot that you can't hold it, something could be wrong. Take it to a pro ... so your phone can chill.

FUN FACT

If your cell phone gets soaked, take out the battery and toss the phone in a bag of uncooked rice. The rice may dry and rescue it!

FUN FACT

Though you might've heard that using a cell phone while pumping gas can cause fires ... studies reveal it's probably a myth.

FUN FACT

Cell phones can explode, but not from being in the sun. Blowups are usually caused by a faulty battery, which is thankfully rare.

HYPERSPECTRAL IMAGING

FUN FACT

The Dead Sea Scrolls, ancient biblical manuscripts discovered inside a cave, were covered in dark marks that made sections unreadable—until hyperspectral imaging revealed the text.

WHAT'S BETTER THAN HAVING A PICASSO? How about finding out you've got another Picasso hidden under the first! When artists aren't thrilled with one painting, they sometimes paint over it. Fine artists' canvas is expensive and takes time to cut, stretch, and prepare. But when an artist paints over an original, it is gone forever. At least, that used to be true, before hyperspectral imaging. Researchers recently used this tech to reveal a whole hidden work beneath Picasso's painting "Mother and Child by the Sea." While a regular camera picks up only the rainbow spectrum we can see, hyperspectral imaging detects reflections of many, many more kinds of light, including ones our eyes can't see. This technique has been used to reveal painted-over images on cave walls and text that's been blotted out by ancient stains like mold or burn marks.

NERD ALERT: SUPER SMARTS

HYPERSPECTRAL IMAGING, also called imaging spectroscopy, has been used for more than 100 years, but not to find hidden pictures. It's been used by scientists to figure out what rocks and minerals are made of. Today, it does this by producing data cubes, which are 3D layers of information.

THIS PORTRAIT WAS FOUND UNDERNEATH THE PAINTING "COUPLE IN A LANDSCAPE" BY 18TH-CENTURY ARTIST THOMAS GAINSBOROUGH.

FUN FACT

Goldfish and bumblebees both see parts of the spectrum humans can't. Goldfish see infrared, and bumblebees see ultraviolet.

BUG BOTS

THEY CAN CRAWL INTO TINY SPACES, LEAP HIGHER THAN BUILDINGS, AND MIGHT EVEN POLLINATE FLOWERS. GET THE BUZZ ON THESE BOTS WITH HIGH-FLYING ABILITIES.

Z6

THIS IS ONE GIANT BUG you'll like having around. Its six insect-inspired legs fold up so it can fit inside a backpack, then unfold to creep through tight spaces and help with search and rescue.

eMOTION BUTTERFLIES

THESE OVERSIZE BUTTERFLIES can flutter along preprogrammed routes, or be controlled by something called indoor GPS. Here's how it works: Infrared cameras send the butterflies' coordinates to a master computer to keep them from colliding. The butterflies could be used to help guide human workers in future factories.

ROBOBEES

INVENTED BY DR. ROBERT WOOD AND HIS TEAM at the Massachusetts Institute of Technology in Cambridge, Massachusetts, U.S.A., these bee bots are about the size of the real thing. They lift off vertically by flapping tiny wings 120 times per second and can hover and steer themselves using sensors and preprogrammed commands. RoboBees may one day help with search and rescue by exploring areas too tight for rescuers to enter. Or they may give actual bees a hand with crop pollination!

COCKROACH ROBOTS

STOMP ON IT, and it keeps on going. This crush-resistant buggy bot is made from piezoelectric material, which means it expands and contracts when electricity runs through it. By adding an elastic polymer to it, its creators made it bend and straighten so it can skitter at 20 body lengths per second. It might someday go into dangerous sites to help find people.

GENETIC SEQUENCING

A SCIENTIST ANALYZES A DNA SEQUENCE.

DNA TESTING KITS PROMISE TO TELL YOU ALL ABOUT YOUR GENES. (Could you have royal ancestry? Or a chocolate ice cream–loving gene?) These genetic kits are limited versions of what a lab can do with full genetic sequencing. The kits use genotyping, which looks for pre-chosen genetic types called variants. But if you have a surprise gene for, say, x-ray vision, a kit won't find it. Genetic sequencing essentially means separating your DNA into strands, then "reading" differences between them. This can be done through an enzyme reaction that copies the strands to reveal their patterns. Another, newer method is nanopore sequencing. It feeds DNA through tiny electrified holes and measures electrical current changes when the DNA passes through. Each DNA molecule generates a unique current change, and that shows which molecule is which.

FUN FACT

The human genome was first fully sequenced by the international Human Genome Project in 2003. Back then, getting your genome sequenced cost about $100 million. Today, it can be done for about $1,000.

FUN FACT

Gene testing kits for dogs can help you learn your pooch's lineage.

NERD ALERT: DO-GOOD GEAR

CRISPR IS A RECENTLY DEVELOPED SHORTCUT to editing people's genes. Short for CRISPR-Cas9, an enzyme that cuts through DNA, CRISPR could improve organ transplants and help develop crops resistant to disease and mold.

WHAT'S YOUR ROBOT ANIMAL SIDEKICK?

These high-tech critter-inspired companions have got your back! Which one is right for you?

START HERE

WOULD YOU RATHER A SIDEKICK THAT'S COOL OR CUTE?

Cutting-edge and supercool. → **WOULD YOU RATHER BE ABLE TO FLOAT OR RUN SUPER FAST?**

- Gravity? No thanks. I'll float.
- Give me speed!

Absolutely adorable! → **WHAT'S YOUR SIDEKICK'S SPECIAL SKILL?**

- It comforts me when I'm sad.
- It knows how to have fun!

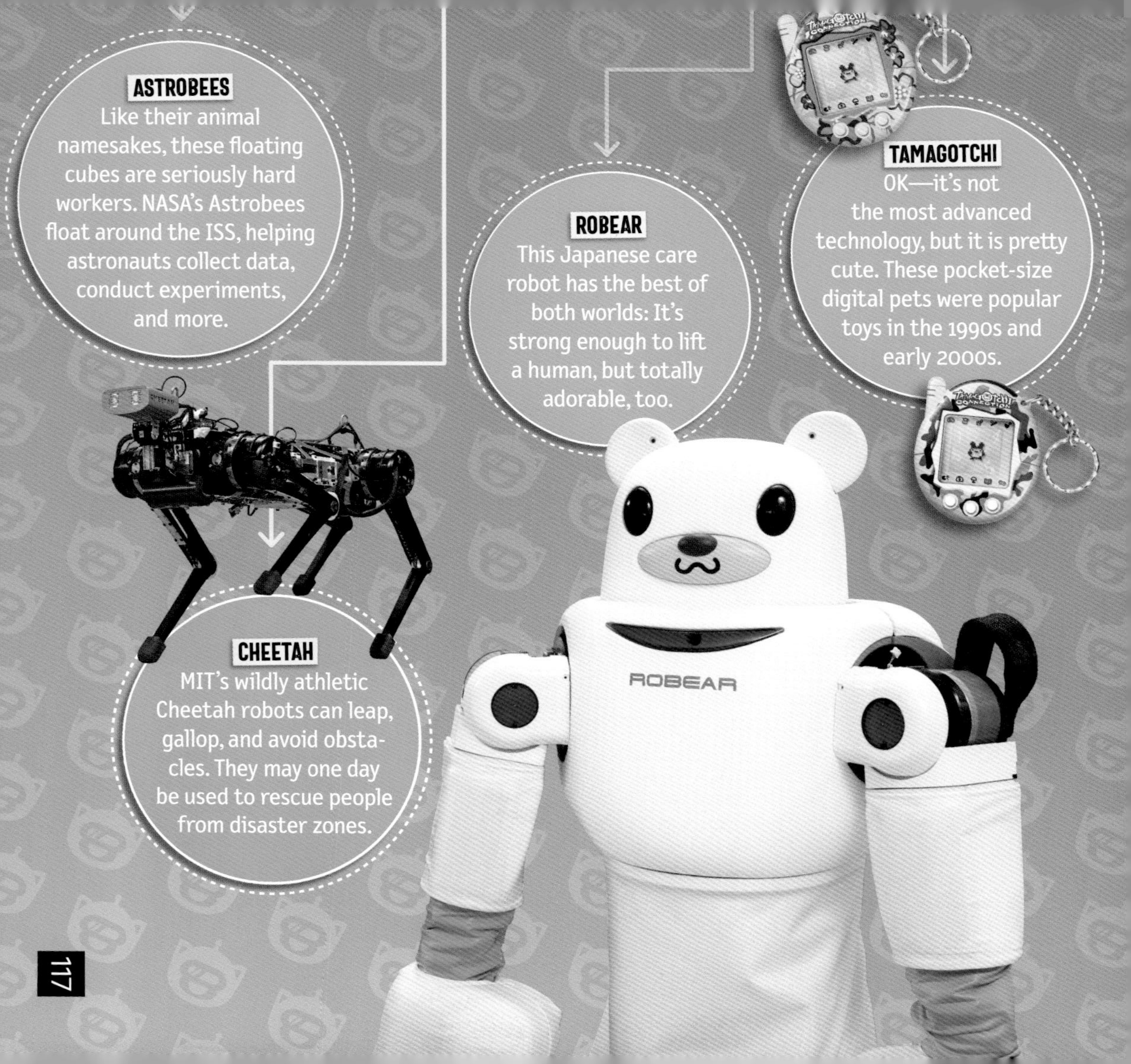

ASTROBEES
Like their animal namesakes, these floating cubes are seriously hard workers. NASA's Astrobees float around the ISS, helping astronauts collect data, conduct experiments, and more.

ROBEAR
This Japanese care robot has the best of both worlds: It's strong enough to lift a human, but totally adorable, too.

TAMAGOTCHI
OK—it's not the most advanced technology, but it is pretty cute. These pocket-size digital pets were popular toys in the 1990s and early 2000s.

CHEETAH
MIT's wildly athletic Cheetah robots can leap, gallop, and avoid obstacles. They may one day be used to rescue people from disaster zones.

GPS

YOU'RE WALKING TO A FRIEND'S HOUSE, when you reach a T in the road. Should you go left or right? You click Google Maps, then wave your phone to catch a signal. In North America, that signal comes from more than two dozen GPS (global positioning system) satellites that the U.S. Department of Defense originally designed for the military. Based on how long it takes your GPS to get a fix on a satellite's signal, your device can figure out how far away it is. Once it's got a fix on four satellites, your device can tell precisely where you are. The satellites send out info called an almanac and an ephemeris. The almanac tells which satellites are in view. The ephemeris tells where each satellite is.

FUN FACT

Different countries have different GPS systems. Russia has GLONASS (GLObal NAvigation Satellite System), and Europe uses Galileo; China has BeiDou.

NERD ALERT: Everyday Egghead

ROOFS AND WALLS WEAKEN AND SCATTER SATELLITE SIGNALS, making GPS less reliable. So engineers have been working on something called indoor positioning. Some systems use strong beacons to make satellite signals penetrate walls. Others place special, flickering lights or even magnets throughout a building. These communicate with your phone to tell it where you are.

FUN FACT

Each GPS satellite weighs around 2,000 pounds (900 kg) and is about 17 feet (5 m) across yet transmits 50 watts or less (similar to a lightbulb).

WHAT DO MODERN SATELLITES HAVE TO DO WITH LONG-LOST CIVILIZATIONS? FIND OUT FROM THE NERD OF NOTE ON THE NEXT PAGE.

NERD OF NOTE:
SARAH PARCAK

Sarah Parcak has her head in the clouds. Scratch that. Her head is higher than that because she's a space archaeologist!

No, Sarah doesn't look for artifacts on other planets. She discovers archaeological sites on Earth that would be hard to see from the ground—for instance, sites in thick jungles or hard-to-access, mountainous regions—and she does it by using satellite images.

The satellites give her a bird's-eye view from nearly 400 miles (644 km) up. Using them, Sarah has found thousands of potential ancient settlements and tombs, and even Egyptian pyramids. Some satellites offer infrared imagery, which can show hidden clues—plants with short roots, for instance, which tip off Sarah that they might be growing over buried structures. But scrutinizing millions of satellite pics can really tire your eyes. So Sarah got the idea to crowdsource—that is, get help from a lot of people—on GlobalXPlorer.org. Once six viewers vote that a pic contains something interesting, it's sent to the pros to check out. How good are the chances of making an Indiana Jones–style discovery? Pretty good, actually. "Literally everywhere we look, we don't know what we're going to find," says Sarah. "There are millions of undiscovered archaeological sites around the world."

"Using these new technologies, we have a real chance to protect and preserve these sites for future generations."

SARAH PARCAK, space archaeologist

Sarah excavates a site in Tanis, Egypt, that she had originally mapped using satellites.

HUBBLE SPACE TELESCOPE

IT'S AS BIG AS A SCHOOL BUS, weighs as much as two elephants, and can search the universe for inhabited alien planets. It's the Hubble Space Telescope, orbiting 340 miles (547 km) above Earth! The giant, silvery tube takes pictures of what it spies in the sky, such as galaxies billions of light-years away and stars being born. Launched in 1990, Hubble uses a camera to detect near-ultraviolet, visible, and near-infrared light. By analyzing the light it receives, the telescope learns about space objects, helped determine the age of the universe (14 billion years!), and detects black holes. It's even found signs of water on other planets, but no alien life has been found ... yet.

THIS IMAGE OF THE HELIX NEBULA WAS TAKEN BY THE HUBBLE SPACE TELESCOPE.

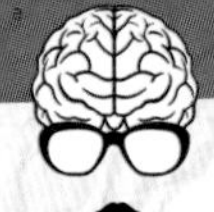

NERD ALERT: SUPER SMARTS

HUBBLE MIGHT NOT ONLY GIVE US GLIMPSES OF THE UNIVERSE but also show us how it all works. Its images have helped reveal dark energy—a force that pushes the universe to expand faster all the time. We know that the universe is 72 percent dark energy, yet exactly what that energy is made of remains a mystery.

DISTANT GALAXIES

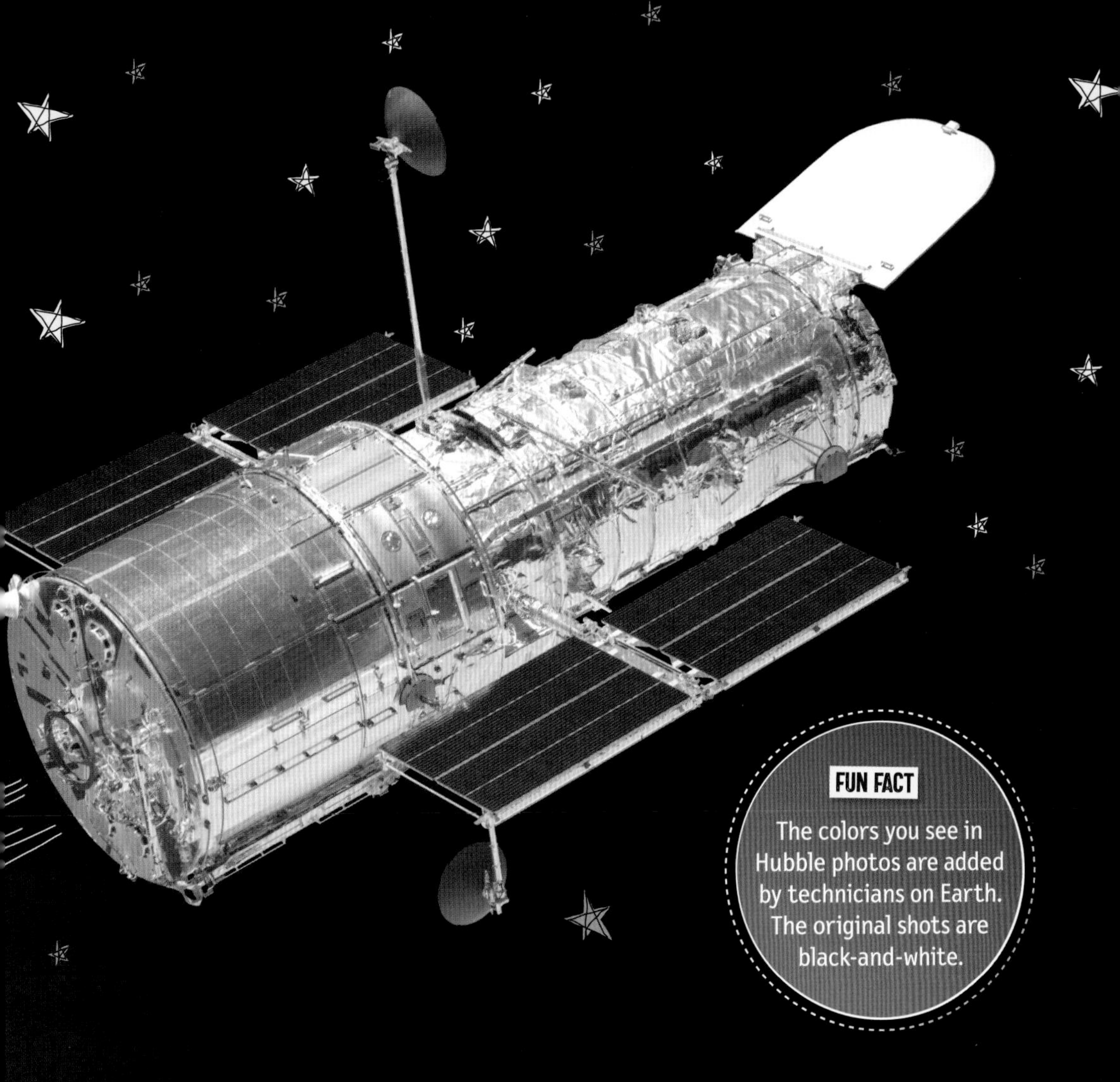

FUN FACT

The colors you see in Hubble photos are added by technicians on Earth. The original shots are black-and-white.

TECHIE TAILORING

IF YOU THINK TECHIE CLOTHING HAS TO HAVE BLINKING LIGHTS AND BUILT-IN CHARGING PORTS, THINK AGAIN. SOME OF THE MOST AMAZING FASHION TECH IS WOVEN RIGHT INTO THE CLOTH.

CONDUCTIVE CLOTHING

THESE CLOTHES MIGHT LOOK OLD-SCHOOL, but don't be fooled. Jacquard by Google jackets have conductive threads woven into them to connect you to your smart device via Bluetooth. Just brush or tap a button on the sleeve cuff to change songs or take a picture. Now that's fast fashion.

FRESH-SQUEEZED STYLE

THE HEALTHY FIBER IN ORANGES turns out to be good for more than drinking. You can wear it! Orange Fiber is an Italian company making textiles from citrus juice by-products (the scraps that usually get thrown away). Cellulose is extracted from the trash-bound peels and fibers and turned into silky yarn. No word on how it tastes.

MODERN MERMAID

COULD YOUR CLOTHES ACT LIKE VITAMINS? That's what SeaCell, a fiber made using brown algae, aims to do. It's made by adding seaweed to fabric, and its makers claim that as you wear it, it'll transfer nutrients from the fabric to your skin, to make it healthier. Maybe, maybe not, since it depends on how much seaweed is in the particular fabric you buy, and there isn't an easy way to check that.

SILKY MILKY

INVENTED IN ITALY BACK IN THE 1930s, milk protein fabric was a hit ... until the 1940s, when World War II food rationing mostly stopped production. Now the fabric is back and still made from casein, aka milk protein. It's warm like wool, but itch-free on skin.

LAB-GROWN MEAT

THIS BURGER didn't start out as a cow. But it's not a veggie patty, either. It's real meat, harvested from cells grown in a petri dish. Besides beef, there's also lab-grown chicken, pork, and seafood. Lab-grown meats start with animal muscle stem cells taken from a living animal under anesthesia. These cells are multiplied in the lab. A single sample can be enough to produce 20,000 pounds (9,070 kg) of meat—while emitting far fewer greenhouse gases and using one-hundredth of the land and a fraction of the water it takes to farm beef. Still, lab-grown meat starts with animals, and the process is expensive. That said, it's a sustainable way to make a whole lot of burgers.

NERD ALERT: DO-GOOD GEAR

SCIENTISTS HOPE THAT LAB-GROWN MEAT will not only reduce the number of beef cattle but also freshen the atmosphere. How? Methane is causing about one-quarter of global warming, and a major source of it is ... cow burps.

FUN FACT

The first lab-grown burger cost about $275,000 to produce in 2013. By 2030, lab-grown meat could come down to about $1.10 per patty.

GMOs

PEOPLE HAVE BEEN TINKERING WITH THEIR FOOD FOR A LONG TIME. Carrots used to be purple and yellow until people bred them to become orange. And the clementines that a lot of us toss into our lunch bags? They're genetic hybrids, made by humans crossing oranges with mandarins. GMOs (genetically modified organisms) are a lot like carrots or clementines, just way more high-tech. Instead of cross-pollinating two different plants, which mixes all the genes in those two plants, scientists insert just one or two genes into a single plant cell. Once that cell divides, the gene shows up in all the new cells. GMOs can make plants bug-resistant, weather-resistant, and even nutrient-enriched.

FUN FACT

The first GMO crop came to supermarkets in 1994. It was the Flavr Savr tomato, which was engineered to ripen without getting squishy.

NERD ALERT: DO-GOOD GEAR

TESTS SHOW THAT GMOS ARE NO LESS SAFE to eat than other produce, but some people don't like the idea that they can be bred to survive certain pesticides—making it more likely that farmers will spray these possibly toxic substances. Others argue that vitamin-fortified GMOs could reduce world hunger. Ultimately, whether our food changes naturally or labs do the work, the future is bound to taste different.

SCIENTISTS HAVE CREATED GOLDEN BANANAS (LEFT) THAT ARE PACKED WITH VITAMIN A.

FUN FACT

A bug called the Asian citrus psyllid has been wiping out orange trees. Scientists are fighting back by engineering oranges with an insect-resistant protein that comes from spinach.

iMAGINE THIS

IF TECH HAD A HOUSE PARTY ...

WHEN YOU STEP OUT OF THE HOUSE, DO YOU THINK YOUR TECH JUST SITS THERE? WHAT IF YOUR FAMILY'S TABLET, ROOMBA, KEURIG, CAT LASER BOT, AND MORE DECIDED TO THROW A HOUSE PARTY? HERE'S HOW THEY MIGHT BOOGIE DOWN.

ALEXA

PARTY TRICK: Ordering the decorations and DJ'ing at the same time

SONG CHOICE: "Say My Name" by Destiny's Child

QUOTE: "Who's up for trivia?"

TALENT: Keeping the conversation going

NEST THERMOSTAT

PARTY TRICK: Warming up the crowd

SONG CHOICE: "Burning Down the House" by Talking Heads

QUOTE: "Is it hot in here, or is it just me?"

TALENT: Making things chill

ROOMBA

PARTY TRICK: Freestyling

SONG CHOICE: "Mr. Roboto" by Styx

QUOTE: "Watch my smooth moves!"

TALENT: Perfect post-party cleanup

KEURIG

PARTY TRICK: Guessing your favorite drink

SONG CHOICE: "Java Jive" by The Manhattan Transfer

QUOTE: "What's the buzz?"

TALENT: Staying up all night

IN-FLIGHT MEALS

WHETHER YOU LIKE YOUR IN-FLIGHT MEAL OR NOT, you should know that serious tech went into creating it. Once the meals are prepped, chefs use a superfast "blast chiller" that cools them to just above freezing. This cuts down on ice crystals forming and breaking food's structural walls, so your meal isn't mushy when it's reheated on board. But that's not all. At 30,000 feet (9,144 m) inside a pressurized cabin, your taste buds are numbed, kind of like when you have a cold. And—believe it or not—constant engine noise changes how you pick up flavors, too. You'll taste up to 20 percent less of the salt and sugar in your meal. So food scientists have cooked up solutions. You'll find lots of savory flavors like tomato and Parmesan cheese, which fare better in the air. Zesty sauces also keep foods from drying out at high altitudes. Now your mystery meal isn't such a mystery!

NERD ALERT: Everyday Egghead

SOME AIRLINE KITCHENS USE A MACHINE called a hydro processor. It uses a computer and a high-pressure stream of water to slice cake, so each piece comes out exactly the same size, with few crumbs.

FUN FACT

Tomato juice is an in-flight favorite for scientific reasons. Its flavor tastes fruitier to our taste buds in the dry, high altitude, and its savory taste becomes extra satisfying.

FUN FACT

Pilots and copilots often eat different meals just in case one meal is poisoned, and the other pilot needs to land the plane!

PET MICROCHIPS

IF YOUR ANIMAL EVER GETS LOST, could a supermarket-style scanner help bring it home? Yes, if your pet is microchipped. A microchip is a miniature radio transmitter inside a capsule about the size of a grain of rice. A veterinarian injects the microchip below your pet's skin, usually between the shoulder blades. Your pet can't feel it once it's in, and the chip doesn't transmit or receive any info until it's scanned. Once it is, the chip beams out a 15-digit number that the scanner reads. This number corresponds to your home info in an online database. If your pet is picked up by almost any shelter or vet's office, they'll scan your animal and see these deets. And if you ever move, the microchip's info can be changed on a computer—there's no need to change the chip, which lasts your animal's whole life.

MICROCHIPS CAN FIT A LOT OF TECH IN A SMALL PACKAGE.

NERD ALERT: Everyday Egghead

MICROCHIPS LIKE THE ONES IN PETS, which are called RFID (radio-frequency identification), are turning up in unexpected places. Some people in Sweden have volunteered to be microchipped between their thumb and forefinger. Their chips are activated by readers just like pets' chips are. But theirs open doors, activate photocopiers, and pay for things when they wave!

FUN FACT

In 2008, a cat named George was reunited with his owners thanks to a microchip ... 13 years after he was lost!

FUN FACT

Lost cats are almost 20 times more likely to be reunited with their families when they're microchipped.

GETTING ARTISTECH

TECHNOLOGY CAN CREATE NEW STYLES OF ART. CONSIDER THESE DIGITAL MASTERPIECES.

LIGHT SHOW

A COMPANY CALLED UMBRELLIUM used lasers to create an art piece titled "Assemblance." By holding hands and moving together, viewers could make the lights into a spectrum of 3D sculptures.

STINKY STYLE

MOSCOW ARTIST DMITRY MOROZOV makes vibrant digital pictures using pollution. A device he created (with a fake nose attached) "sniffs" out pollution with sensors, then connects to a computer that translates the data into brilliant colors.

WILD WINDOWS

PICTURE A HOUSE OF THESE: Artist Eric Standley prints over-the-top paper creations that look like stained glass windows. He programs the designs into a computer that cuts hundreds of images with lasers. Then he stacks them to make extreme patterns.

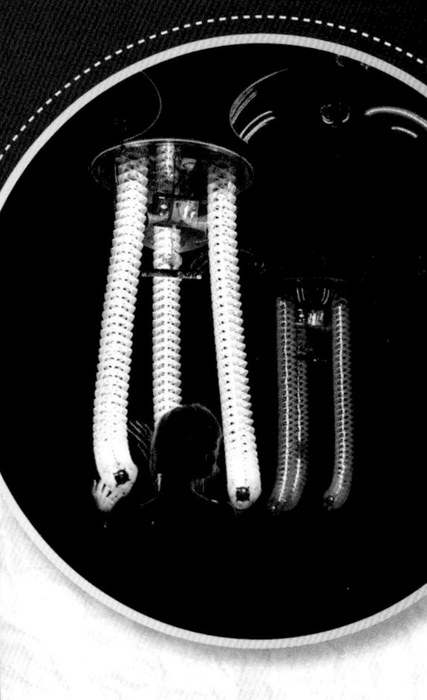

ROBO-ZOO

PLEASE TOUCH THIS ART! Petting Zoo by Minimaforms is a series of hanging, snakelike robots that change colors and move when visitors pet, touch, and talk to them.

ROBO HIGH
CLASS SUPERLATIVES

Most On the Ball

AS THE FIRST EVER ROBOT DESIGNED TO TRAIN HUMAN ATHLETES, Forpheus looks like a big alien that plays a mean game of ... table tennis. The Japanese robot's cameras-for-eyes and motion sensors track your body, your paddle, and the ball, then use that info to predict where the little sphere will fly next. Forpheus also figures out how good you are and adjusts its game to your skill level. To help you improve, it compares your moves to the experts' and offers pro tips. A true coach, this bot would be proud to have you beat it at its own game.

See you in summer sports!

—Xingzhe No. 1

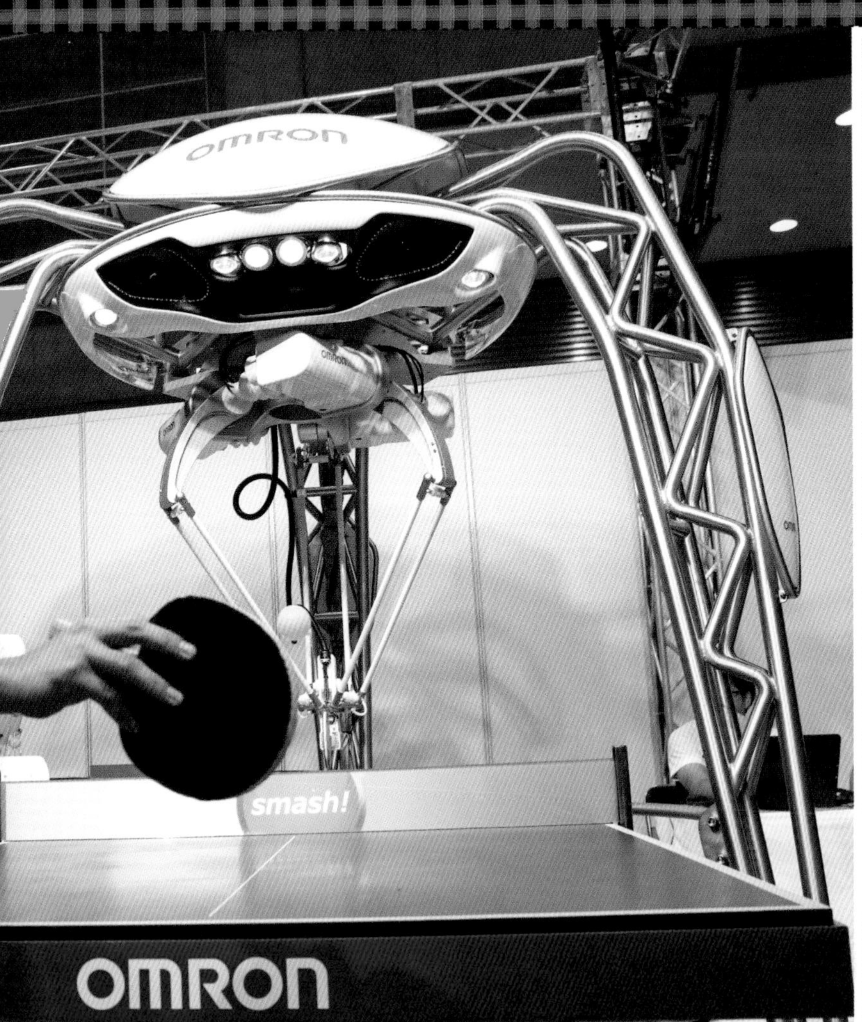

MONONOFU, **page 12**

Xingzhe No. 1, **page 56**

Alpha 1S, **page 96**

Tradinno, **page 174**

JUSTIN DIFUTURE, MD

PAINLESS SHOTS, MICROSCOPIC SPONGES, AND PERSONALIZED VACCINES MIGHT ALL BE PART OF MEDICINE'S TECHIE FUTURE. CHECK OUT THESE CUTTING-EDGE HEALTH INNOVATIONS.

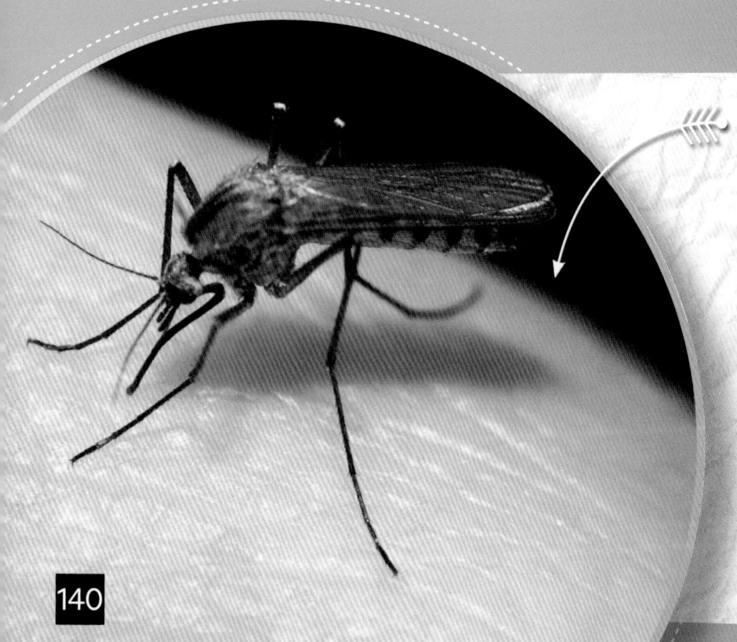

MOSQUITO INJECTIONS

WHEN A MOSQUITO POKES its needlelike proboscis into your skin, it doesn't hurt. Researchers at Ohio State University in Columbus, Ohio, U.S.A., are designing a micro-needle that would be just as painless. Its secrets: a numbing agent, a zigzag-shaped needle that vibrates, and both soft and hard parts that enter skin more easily than today's needles.

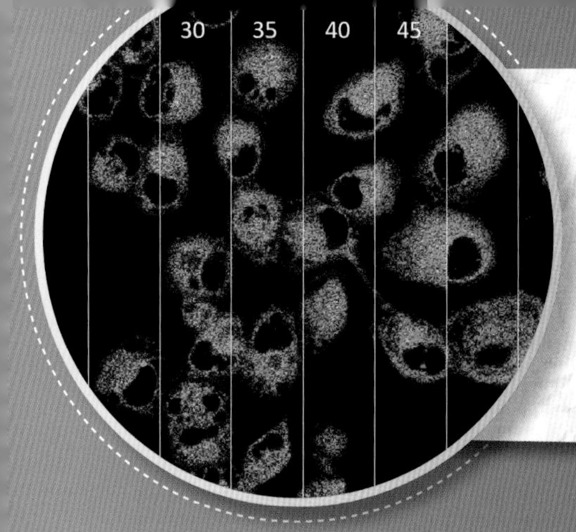

NANO-THERMOMETER

HOW DO YOU TAKE A SINGLE CELL'S TEMPERATURE? With a nano-thermometer. Chemists at Rice University in Houston, Texas, U.S.A., made theirs from molecules of boron dipyrromethene (nickname: BODIPY). Inside a cell, BODIPY spins and lights up as it gets hotter. Once it's hot enough, its light turns off. Under a microscope, researchers can see which cells are hottest—and more likely to be cancerous.

SNIFFPHONE

FINDING CANCER EARLY is one of the best ways to stop it. So researchers thought: What if you could sniff out cancer by breathing into your phone? The SniffPhone is a compact handheld device that uses sensors to detect certain chemicals generated by gastric cancer. If something is found, the phone alerts medical pros.

NANOSPONGES

THESE MICROSCOPIC CLUSTERS of polymers—essentially, itty-bitty sponges—could be injected into your bloodstream to soak up bad bacteria, venom, and other toxins and wash them out of your body. The mini sponges are "disguised" inside blood cell membranes, so the body gives them VIP access to every part of your system, where they sop up gunk as they go.

HAVE QUESTIONS ABOUT MEDICAL TECH FOR THE HERE AND NOW? CHECK OUT THE NEXT PAGE.

FAQ DO I NEED A FLU SHOT EVERY YEAR?

You got a flu shot last year. And the year before that. Seriously, is your doc jabbing you for kicks?

Nope. To get why flu is such a big deal, it pays to know a little history. Back in 1918, there was a worldwide flu epidemic. People thought they were just getting a cold, but then it got serious.

Hours after catching it, people were dying. Within two years, up to 50 million people had died, making some experts believe that it might be the end of humanity. There were no flu vaccines back then. By the end of 1919, whoever had survived infection was immune, and that version of the virus hasn't shown up again. But every time flu viruses multiply, their genes change a little, thanks to what is called antigenic drift. That's why, every year, there's a new flu to fight—making vaccines one of our best defenses.

Still, if you're one of those people who never gets the flu, why do you need a shot? You could actually be infected without symptoms. That's no big deal for you, but it's potentially dangerous for people with weakened immune systems, such as babies and the elderly. Coughing, sneezing, or even talking within six feet (1.8 m) of them can spread the virus. So getting yourself protected means protecting someone else, too.

FUN FACT

Researchers are working on a universal flu vaccine that would be the only flu shot you'd need for the rest of your life.

FUN FACT

In 2005, under extreme biosecurity, Dr. Terrence Tumpey re-created the 1918 flu virus to essentially hack its deadly genes to learn how to stop similar viruses in the future.

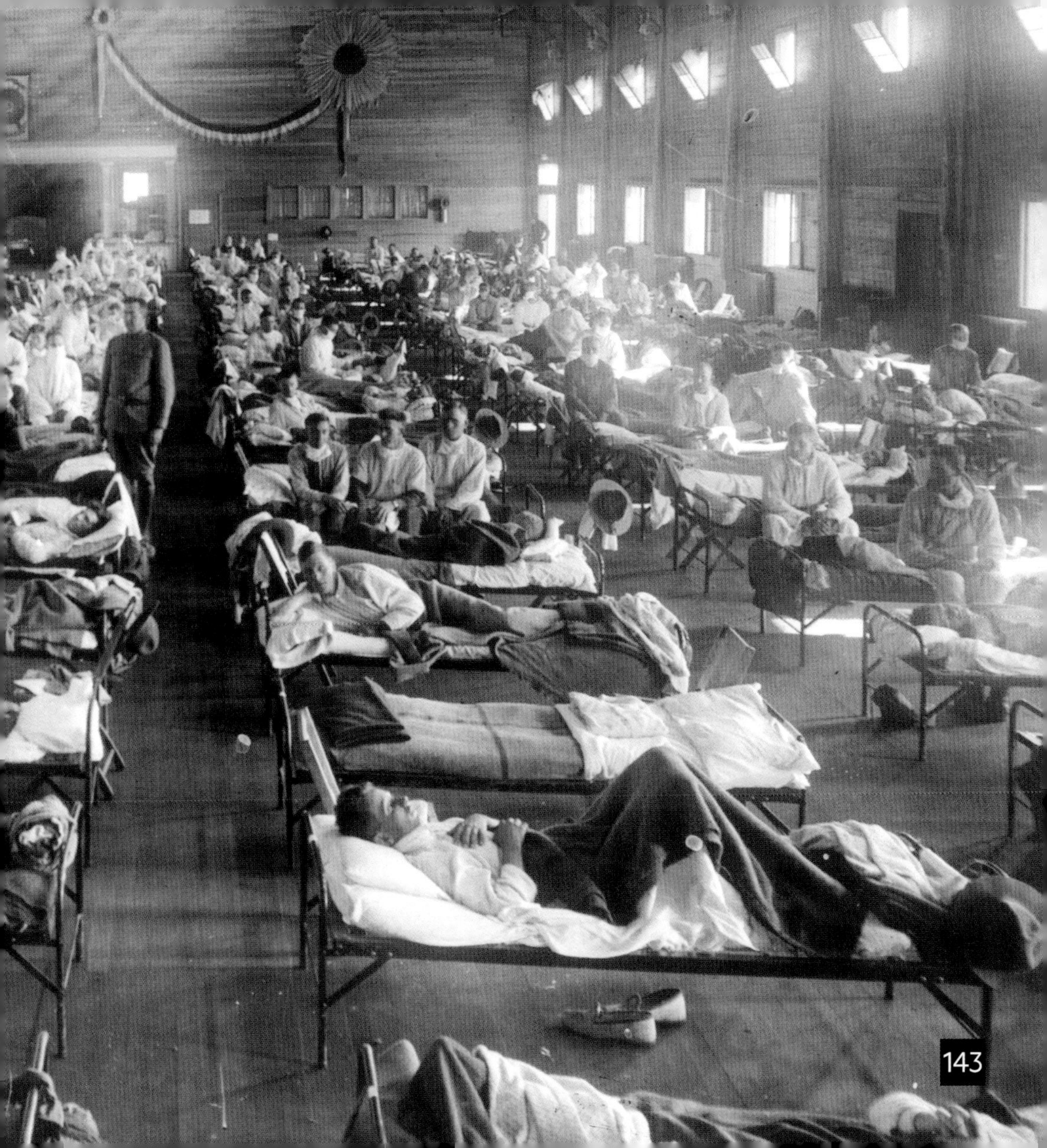

NEURAL MICROSENSORS

YOU MIGHT CALL YOURSELF A TECH HEAD, but would you put actual tech inside your head? Some people are already doing it with neural microsensors—tiny sensors implanted in their brains to enhance their abilities, bring back abilities they've lost, or add ones they never had. Would you like to move a computer cursor just by thinking about it? While it would be a cool trick, it could also help people who can't use their limbs because of a spinal cord injury or stroke. A system called BrainGate makes it happen by using wireless microelectrode implants. When users imagine moving their paralyzed limbs, the implants send neural signals to a computer instead, and move the cursor. This tech has also helped people control bionic limbs the way they would control natural limbs. Maybe one day, everyone will get where they're going just by thinking about it.

BRAINGATE SENSOR

NERD ALERT: DO-GOOD GEAR

NEURAL MICROSENSORS are getting smaller and more powerful. An international team is working on "neurograins," wireless brain implants the size of a grain of salt. The plan is for them to record what the brain is doing and send pulses to help bring back lost functions.

FUN FACT

One man, Kevin Warwick, volunteered to have a BrainGate implanted to help test the tech. Now he can turn lights on, set an alarm, and operate computers by opening and closing his hand.

PROGRAMMABLE MATERIALS

YOU OPEN A FLAT BOX with a picture of a table on it, then jump back. The table starts expanding and assembling itself! Anyone who's ever struggled with the pieces inside a DIY kit will appreciate programmable materials—that is, stuff "programmed" to change without any human help. Picture replacement body parts programmed to "grow" with you, or pipes that expand during flooding and bend during earthquakes. And how about braces that adjust without an orthodontist's help? The materials change after being triggered by certain stimuli—temperature, water, electricity, light, or pressure.

THIS FLAT SHAPE DOESN'T LOOK LIKE MUCH ...

... UNTIL IT TRANSFORMS!

NERD ALERT: SCI-FI COOL

SOME PROGRAMMABLE MATERIALS COULD BE CREATED THROUGH 4D PRINTING. 4D printing is a lot like 3D printing, a process in which whole objects are created by putting layers of material on top of each other. But, in the case of 4D printing, the objects would also be able to change shape. That's where the fourth dimension comes in: time.

FUN FACT

Researchers in the Netherlands have studied origami to create programmable crease patterns called an origami alphabet. Some uses for the tech include foldable robots and solar panels.

iMAGINE THIS

SUPERSTRONG MATERIALS

KNOCK DOWN

IMAGINE IF SOME OF THE WORLD'S STRONGEST MATERIALS WENT HEAD-TO-HEAD. WHO'D WIN? HERE ARE THE CONTENDERS.

COMPOSITE METAL FOAM (CMF)
CHARACTERISTIC:
Filled with impact-absorbing micro holes
SPECIAL POWER:
Steel CMFs can stop a flying bullet at half the weight of regular steel.
WEAKNESS:
It's really heavy.

HAGFISH SLIME

CHARACTERISTIC:

Traps predators in goo

SPECIAL POWER:

Each thread expands 10,000 times when released into water and is five times stronger than steel.

WEAKNESS:

It can be scraped off by wriggling.

THE **U.S. NAVY** STUDIED **HAGFISH SLIME** TO CREATE TECHNOLOGY THAT HELPS **PROTECT DIVERS** WHILE UNDERWATER.

DRAGON SILK

CHARACTERISTIC:

Packed with spiderweb protein

SPECIAL POWER:

Made by genetically enhanced silkworms, it's up to 37 percent stretchier than Kevlar (the go-to for bullet-proof vests).

WEAKNESS:

It's only two-thirds as strong as Kevlar.

DRAGON SILK THREAD

HEALING VENOMS

YOU MAY HAVE HEARD THAT WHAT DOESN'T KILL YOU MAKES YOU STRONGER. WELL, THAT'S THE SURPRISING TRUTH BEHIND THESE VENOMS, WHICH CAN HEAL WITH A LITTLE HELP FROM MEDICAL TECH.

CONE SNAIL

THINK SNAILS ARE CUTE? Not this one. Its paralyzing bite means the end for the small fish it eats. Fortunately for humans, its venom can be made into an extremely strong painkiller.

SEA ANEMONE

ITS PASTEL swaying tentacles look pretty, but they're packed with neurotoxins that stop predators and paralyze prey. Handily, these poisons also obliterate human lung and breast cancer cells in lab tests.

BLACK MAMBA

THE BLACK MAMBA has the deadliest bite of any snake on Earth. That said, its would-be fatal venom contains pain-killing proteins called mambalgins that are as strong as morphine but aren't addictive.

DIPLOCENTRUS MELICI SCORPION

THIS DESERT DWELLER can paralyze and kill victims with its stinger. But compounds in its venom have also been shown to kill drug-resistant tuberculosis and staphylococcus bacteria.

JET PACKS

WISH YOU COULD FLY LIKE IRON MAN? You can, with a jet pack. Some inventors—many of whom are as wealthy and nerdy as the fictional Tony Stark—have made flying dreams come true. Back in 1984, test pilot Bill Suitor flew into the opening ceremony of the Olympic Games using a jet pack. Suitor said it felt like "trying to stand on a beach ball in a swimming pool." Things have accelerated since then. In 2015, David Mayman and Nelson Tyler created a pack using jets that tilt to steer. Athlete Richard Browning created a 3D-printed aluminum shoulder harness with two mini jet engines on each arm and one attached to the back. All the suits fire off significantly more horsepower than most race cars, and some can soar higher than 12,000 feet (3,660 m) going as fast as 150 miles an hour (241 km/h).

FUN FACT

Yves Rossy, also known as Jetman, created a jet pack with wings that can take him as high as planes. To fly it, he has to register himself as an aircraft!

NERD ALERT: SCI-FI COOL

JET PACKS USED TO RUN ON HYDROGEN PEROXIDE, but it burned so fast that they stayed up for only about 30 seconds. Today, they use purified kerosene and can fly for around 10 minutes, depending on how fast you fly.

FUN FACT

One of the first jet packs in pop culture belonged to comic book hero Buck Rogers.

FLYING CARS

YOU DRIVE TO THE AIRPORT AND ONTO THE TARMAC, then press a button. Wings fold out. Your car becomes a plane! You'll likely need both a driver's license and a pilot's license to operate this vehicle, because it really is a car and a plane combined. Most of today's models look kind of like a minivan from the front and a tiny plane from the back, but they've each got a regular steering wheel, take ordinary gasoline, and fit in a home garage. So why don't we see them everywhere? Well, getting approval from the Federal Aviation Administration is complicated—the FAA has to figure out air traffic control for all these flying family cars, after all. Plus, they're noisy. And there's the price, hovering around $300,000. In spite of everything, though, experts think we could see these things take flight within a decade.

FUN FACT

You can see some of the world's first flying cars on display at the National Air and Space Museum in Washington, D.C.

FUN FACT

The Mizar was a regular Ford Pinto that could be connected to wings at the airport. It flew, but the project was ultimately considered a failure after one failed flight.

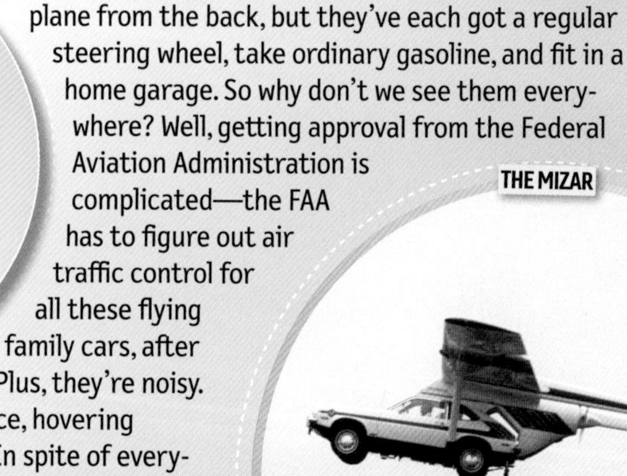

THE MIZAR

NERD ALERT: SCI-FI COOL

THE FIRST CERTIFIED PLANE that could also be driven on U.S. highways was the Taylor Aerocar, created in the 1940s. It worked but was too pricey to produce. Since then, inventors have tried to make science fiction a real-life fact—and soon, we may actually get there.

GPR

RADAR WORKS BY SENDING OUT HIGH-FREQUENCY RADIO WAVES. When the waves bounce off a surface, the radar system computes how long it takes for them to come back. By combining that info with the direction and frequency of the reflected waves, radar can tell how far away the surface is, how big it is, and how fast it's moving. Ground penetrating radar (GPR) goes a step further: Its lower frequencies penetrate as far as hundreds of feet below Earth's surface. It's handy for finding underground pipes or building foundations. GPR was even used to search for hidden rooms in King Tut's tomb. Researchers were able to peek in without making a single crack or drilling one hole. Unfortunately, no surprise rooms were found. But on the flip side, no priceless artifacts were destroyed while figuring that out!

A GPR DEVICE ON WHEELS

FUN FACT

LIDAR uses pulses of laser light instead of radio waves to map Earth's surfaces, both on land and underwater.

NERD ALERT: DO-GOOD GEAR

GPR HAS BEEN CHANGING ARCHAEOLOGY BY FINDING seriously impressive artifacts without breaking ground. Big finds include the remains of King Richard III under a British parking lot and new-to-us areas of Pompeii, an ancient Roman city buried under volcanic ash by the eruption of Mount Vesuvius in A.D. 79.

USING GPR TECHNOLOGY, ARCHAEOLOGISTS CAN SCAN A WALL IN KING TUT'S TOMB WITHOUT DISTURBING THE SITE.

FUN FACT

It's possible for objects to "hide" from radar. The B2 stealth bomber's sharp angles and metal-coated windows scatter radio waves, which make it hard for radar to detect it.

NERDSVILLE CENTRAL: SPACESHIPTWO

HIGH-FLYING ADVENTURE

OK, SO IT COSTS $450,000 PER TICKET. But what a ride! A true destination vacation, Virgin Galactic's SpaceshipTwo will fly you about 51 miles (82 km) up, up, and away to the edge of space, where you float around gravity free for about five minutes. To get there, the aircraft is carried by WhiteKnightTwo, a four-engine craft that attaches to SpaceshipTwo and looks like a pair of planes surrounding it. Once it's nine miles (15 km) up, WhiteKnightTwo breaks away, and SpaceshipTwo fires its rocket engine to rev up. The rocket shuts down and whoosh, the ship coasts into space! From there, you'll be able to see the curve of our planet and the blackness of space. To come home, the ship's wings and tail rotate upward, turning the ship nose down. The aircraft is nearly ready for commercial tourism. Some VIP passengers get to ride for free—plants, dust, liquids, and gases. They all soar gratis as part of research experiments aboard SpaceshipTwo.

1 Tour the Spaceport. Even without a ticket to fly, you can book a four-hour tour of Spaceport America in New Mexico.

2 Do the VR. On Virgin Galactic's website, you can virtually experience flight.

3 Get spin-ready. Once you've got your ticket, you'll prep for three days. This includes boarding the PHOENIX, a centrifuge that spins you in a simulator cockpit at dizzying speeds.

4 Look outside. SpaceshipTwo is built with as many windows as possible so you can see space from all angles (especially while floating!).

MARS ROVERS

THE MARS HELICOPTER: INGENUITY

THE FIRST OF THESE FIVE ROLLING ROBOTS, Sojourner, landed on the red planet in 1997. Rumbling along at just two hundredths of a mile an hour (.03 km/h), it was the size of a microwave oven, and it sent home about 550 photos while it analyzed Martian dirt and checked out the planet's weather. In 2004, twin golf cart–size rovers Spirit and Opportunity were the first to find evidence that water once flowed on Mars. In 2012, Curiosity was sent to Mars to find out whether the planet ever hosted life—well, microbes, at least. And in 2021, Perseverance, the latest rover, joined Curiosity on Mars, conducting its own search for signs of past life.

CURIOSITY

NERD ALERT: SCI-FI COOL

ORIGINALLY SENT ON A TWO-YEAR-LONG MISSION, Curiosity keeps on going and going. And it's got the incredible tech to get the job done. At about the size of a small car, the rover has 17 cameras that both snap pics and act as its "eyes," a laser to blast through rock samples for study, and a seven-foot (2.1-m) arm that handles tools *and* acts as a selfie stick. So far, it's determined that a salty lake may have existed on the planet about 3.7 billion years ago. It's now looking for ancient signs of life.

FUN FACT

Perseverance has a companion on Mars: a helicopter called Ingenuity! This Mars helicopter is new tech that can explore the red planet from above while rovers explore below.

WICKED RIDES

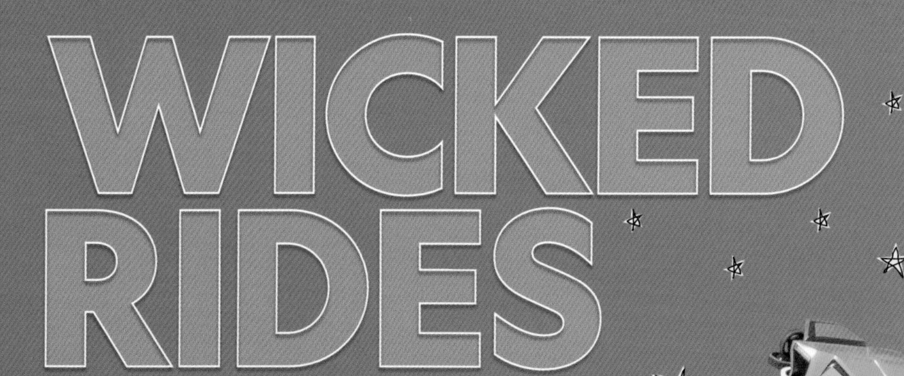

SOME TECH IS JUST FOR FUN ... IF PLUNGING THOUSANDS OF FEET AT HAIR-RAISING SPEEDS IS YOUR IDEA OF FUN! BRACE YOURSELF FOR THE WORLD'S MOST EXTREMELY ENGINEERED ROLLER COASTERS.

SPEED DEMON

RACE CAR–THEMED COASTER FORMULA ROSSA in Abu Dhabi sets the world speed record for roller coasters at 149.1 miles an hour (240 km/h).

SHEER DROP

THE TMNT SHELLRAISER, inside New Jersey's American Dream mall, holds the record for the world's steepest drop, at 121.5 degrees. This means that, instead of falling straight down, it zooms down and angles in. It has a magnetic launch, with the train and the track working like magnets repelling each other. Whoosh!

ROCKET LAUNCH

WHAT DOES IT FEEL LIKE to be shot from a cannon? Do-Dodonpa, a Japanese coaster named after the sound of drum beats, has the highest launch acceleration on Earth. It goes from 0 to 112 miles an hour (180 km/h) in 1.56 seconds by using compressed air. Riders say it sounds explosive!

GREEN MACHINE

IT'S EASY TO IMAGINE THE HULK flinging giant tires around. So it kinda makes sense that the Incredible Hulk Coaster at Universal in Orlando, Florida, U.S.A., is tire-propelled. It works by using motorized tires to push cars up an incline.

THEREMIN

WHAT KIND OF INSTRUMENT can you play without touching it? A theremin! Invented by Léon Theremin in 1920, this electronic device looks like a box with a straight antenna on one side and a looped antenna sticking out the opposite side. Wave your hands near the antennas, and you create a circuit that produces a tone. It sounds like a high-pitched voice or stringed instrument. The sound comes from "heterodyning," or mixing two different, similar frequency signals to create a new signal whose frequency equals the difference between the two. The hand placed near the straight antenna controls the pitch, while the hand held next to the looped antenna controls the volume. Most people say it sounds otherworldly or spooky. After all, it's an instrument nobody plays!

NERD ALERT: VINTAGE VISIONARY

MUSICIAN CLARA ROCKMORE DEVELOPED a special finger technique for playing the theremin. She had to keep the rest of her body extremely still because any movement changes the pitch. And no one could walk by while she was playing because that would mess up the music.

FUN FACT

You can hear theremins in space-themed sci-fi movies, including *The Day the Earth Stood Still*, *It Came From Outer Space*, and *Mars Attacks*.

FUN FACT

The original name for the theremin was the aetherphone.

HUG-A-BOTS

SOME ROBOTS HAVE ONE JOB: TO MAKE YOU LOVE THEM. HERE ARE THE MOST LOVABLE, HUGGABLE BOTS.

AIBO

THIS ROBO-PUP CAN SHAKE ITS HEAD, wag its tail, scratch an ear, and fetch ... but in a lower-key way than a real dog. Using camera eyes, voice recognition, and sensors, Aibo follows its owner and responds to petting.

KIKI

THIS FEISTY LITTLE ROBOT will develop a unique personality based on your behavior. She'll respond to your emotions and be shy around new people, thanks to facial recognition and artificial intelligence.

QOOBO

IS IT A CAT? A DOG? A FOX? It's really a cushion with a tail, so you can decide which pet you'd like this robot to be. Its tail swishes or wags when you pet it, simply communicating love.

LOVOT

THIS FUZZY, slothlike bot has a "sensor horn" that's packed with a 360-degree camera, sound and light sensors, and a thermal camera so it can scan a room to find you. Designed to be relaxing, Lovot has an air circulation system that makes it feel warm like a real animal, plus it responds to your mood by singing.

EXTENDED REALITY

"EXTENDED REALITY"—OR XR— is the term for all the different kinds of virtual reality. Yep, there's more than one! While true virtual reality exists within one artificial landscape—like the 360-degree experience you'd get by looking through VR glasses—augmented reality blends screens with your world. You might know it from Pokémon GO, in which pointing your phone at the real world reveals monsters on your screen. Mixed reality uses goggles to let you interact with the real and virtual worlds at once. You might need to make room for an animated robot in your desk chair, or a realistic pterodactyl perching on your bed. Extended reality theme parks add sets with sound and weather effects that make it seem like you're really in a whole new land.

AUGMENTED REALITY HELPS YOU CATCH POKÉMON IN THE REAL WORLD.

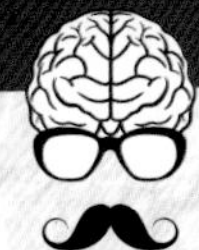

NERD ALERT: SUPER SMARTS

VIRTUAL REALITY CAN NOW GIVE YOU a peek at the past—museums are debuting ways for you to virtually manipulate fragile treasures that usually stay behind ropes. The London Natural History Museum lets visitors "touch" specimens like a pterosaur and a blue whale. Some XR sets add smells and sounds, too!

FUN FACT

A German water park has a VR-enhanced waterslide with waterproof headsets.

VOCALOIDS

VOCALOID SUPERSTAR HATSUNE MIKU "STANDS" ONSTAGE.

WOULD YOU PAY TO SEE POP SINGERS WHO DON'T EXIST? Thousands of people already do! The virtual performers are called Vocaloids (vocal androids). They're synthesized singing voices paired with animation. Onstage, they look like giant holograms, with live bands to back them up. Programmers create Vocaloids by having a live singer record their voice as evenly as possible into a computer. Users then access the bank of sounds to create songs. Each Vocaloid has a unique sound, and you pick the one you want to perform the song you write. When it's finished, you can pair it with a DIY video and release it online, where there are thousands of Vocaloid fans ready to listen.

FUN FACT

The first Vocaloid "star" was Hatsune Miku—she was made to look like a 16-year-old girl with superlong ponytails.

NERD ALERT: SCI-FI COOL

VOCALOIDS APPEAR TO PERFORM ONSTAGE, thanks to a technique called Pepper's Ghost. This works by placing an angled piece of glass in front of a brightly lit stage. In a hidden, darkened area, an actor—or in this case, a projector—waits. When the stage lights go dim and the hidden area is lit up, the projection is cast on a mirror that reflects onto the glass. To the audience, it looks like a hologram!

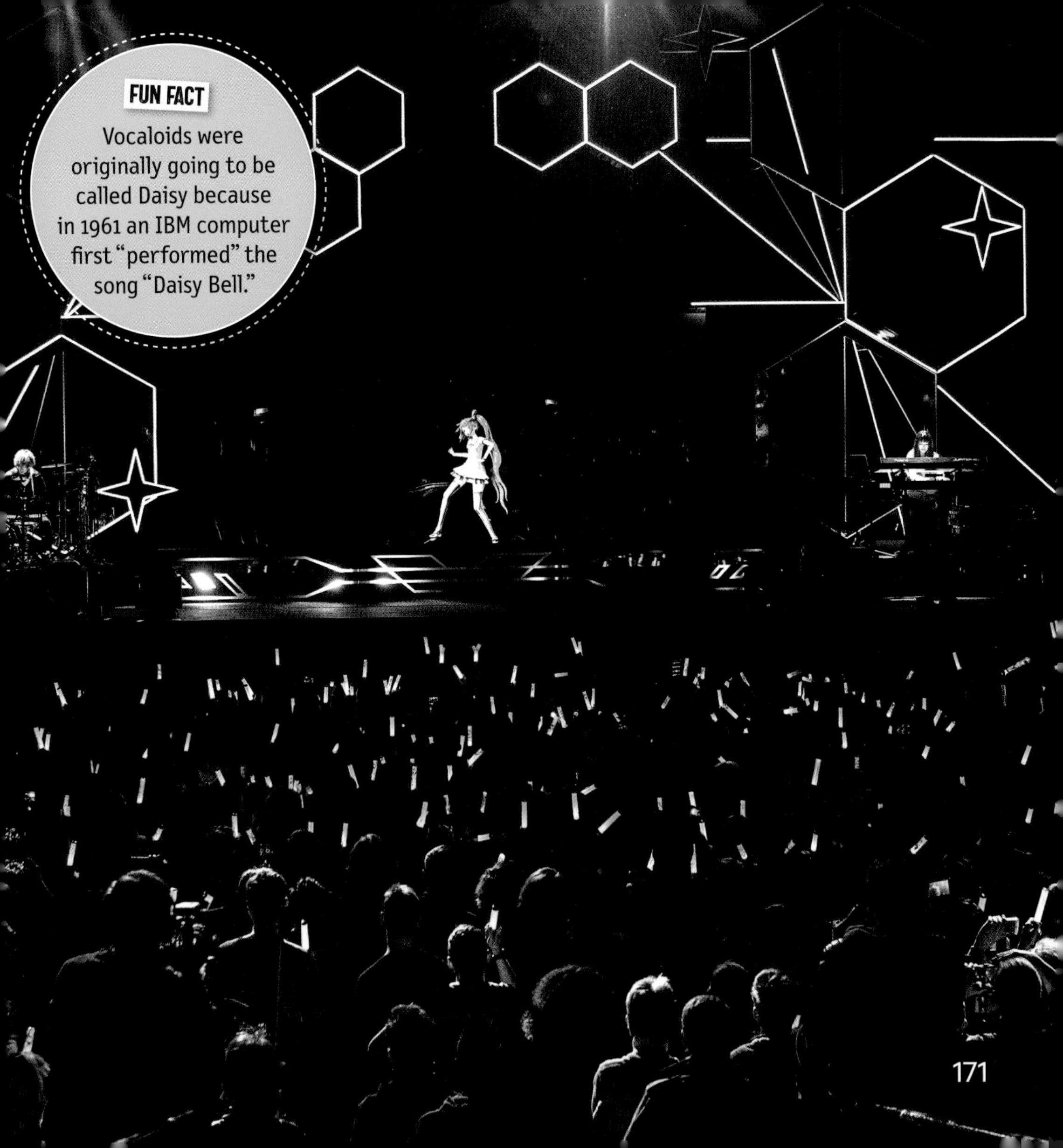

FUN FACT

Vocaloids were originally going to be called Daisy because in 1961 an IBM computer first "performed" the song "Daisy Bell."

WHICH FICTIONAL ROBOT IS MOST LIKE YOU?

YOU PROBABLY HAVE MORE IN COMMON WITH THESE FICTIONAL ROBOTS THAN YOU REALIZE. FIND OUT NOW!

START HERE

WOULD YOU RATHER BE FAMOUS FOR KINDNESS OR STRENGTH?

Strength. → **ARE YOU BRAVE?**

Kindness. → **YOU'VE BEEN CAPTURED BY AN ENEMY! WOULD YOU RATHER TALK YOUR WAY OUT OR USE TOOLS TO ESCAPE?**

Totally. → **ARE YOU BETTER ONE-ON-ONE OR LEADING A TEAM?**

I prefer my awesome debating skills.

I'm a gadget whiz—give me tech and watch me go.

I'm a born leader.

One-on-one is best.

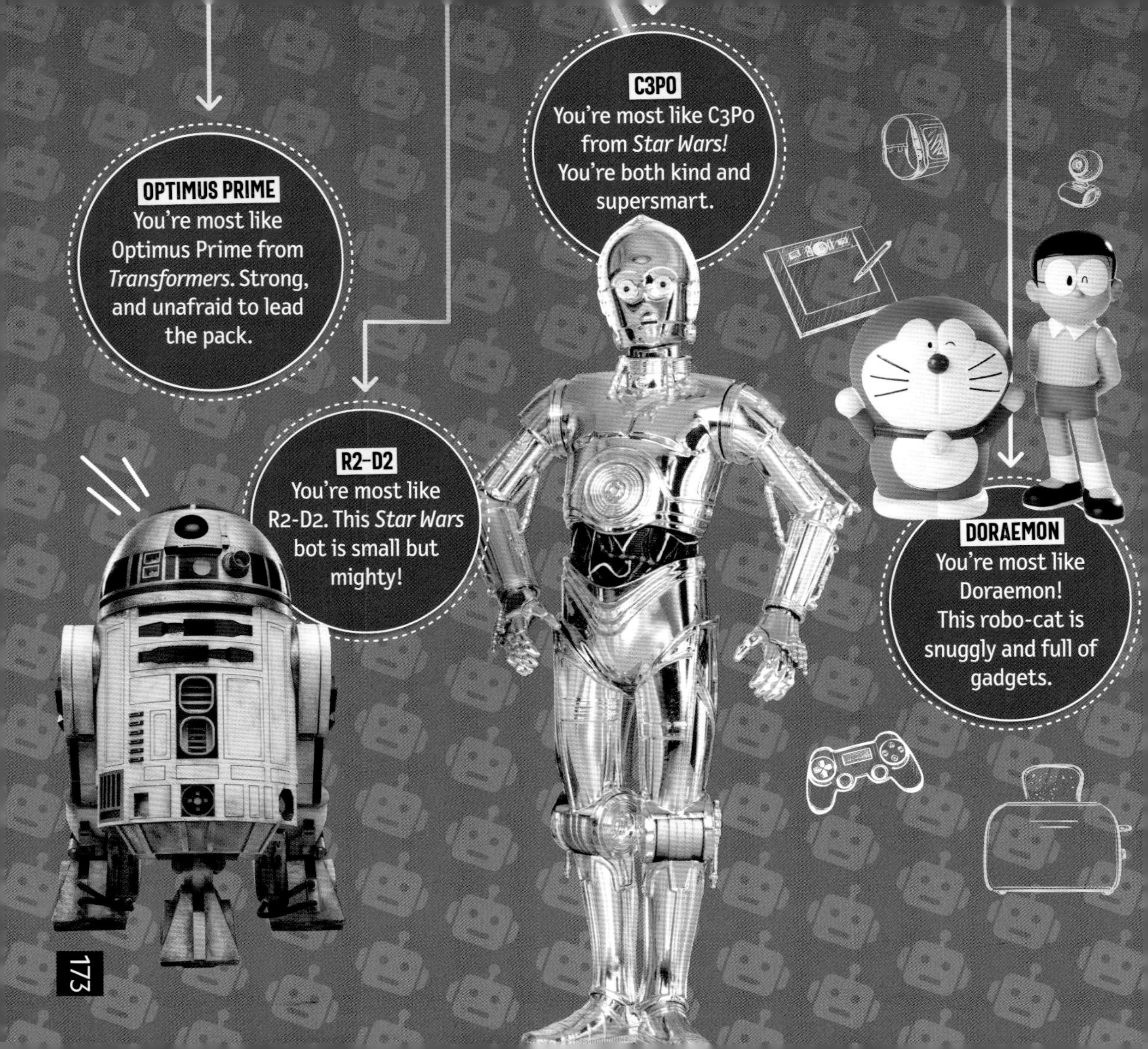
OPTIMUS PRIME
You're most like Optimus Prime from *Transformers*. Strong, and unafraid to lead the pack.
R2-D2
You're most like R2-D2. This *Star Wars* bot is small but mighty!
C3PO
You're most like C3PO from *Star Wars!* You're both kind and supersmart.
DORAEMON
You're most like Doraemon! This robo-cat is snuggly and full of gadgets.

ROBO HIGH
CLASS SUPERLATIVES

Most Likely to Win an Academy Award

IT ROARS, BREATHES SMOKE AND FIRE, AND CAN'T GET ENOUGH OF THE SPOTLIGHT. It's Tradinno, a massive winged dragon robot built to star in the traditional German folk play "Drachenstich." Operated by remote control, Tradinno—whose name is a combo of the words "tradition" and "innovation"—terrorizes the play's human characters with its nearly 52-foot (15.5 m)-long reptilian bod (that's about 12 feet [3.6 m] longer than a *T. rex*). Adding to its performance skills are its abilities to "bleed" fake blood, swish its tail, and contract and expand its pupils. And of course, take a bow.

MONONOFU, **page 12**

Xingzhe No. 1, **page 56**

Alpha 1S, **page 96**

Forpheus, **page 138**

ULTRA ANIMALS

THANKS TO TECHNOLOGY, ANIMALS THAT HAVE LOST LIMBS, FINS, BEAKS, OR FLIPPERS CAN POWER UP WITH PROSTHESES. MEET SOME BIONIC CRITTERS!

POWERED-UP PIG

CHRIS P. BACON WAS BORN WITH paralysis in his back legs. So Florida, U.S.A., veterinarian Len Lucero thought, Why not create a mini wheelchair? He took apart some of his son's old toys to create a harness with wheels. Chris took to it right away, walking and hopping.

TECHY TURTLE

ALLISON THE SEA TURTLE lost three flippers in a predator attack, which meant she could only turn herself in circles in the water. Then a conservationist at the turtle rescue center Sea Turtle, Inc. on South Padre Island, Texas, U.S.A., realized something: If Allison had a rudder, it would stabilize her like a boat. A carbon fiber prosthesis with a finlike rudder was clipped to her shell. Now, Allison can cruise anywhere she wants—and look pretty futuristic doing it.

ALLIGATOR ACTION

WHEN MR. STUBBS LOST HIS TAIL, he couldn't make the signature alligator "swish" that helped him surface from swimming underwater. So a team at the Phoenix Herpetological Society in Scottsdale, Arizona, U.S.A., used a 3D printer to make a new tail just his size. When Mr. Stubbs used the new tail to slap a volunteer, the team knew it was working!

BETTER BEAK

JARY THE HORNBILL HAD CANCER ON HIS BEAK, meaning that part of his beak needed to be removed. Jary's vets at the Jurong Bird Park in Singapore had a solution. They removed the cancer and 3D-printed a prosthetic beak for Jary that was a perfect fit.

PILLCAM

SEEING DEEP INSIDE YOUR SMALL INTESTINE used to mean undergoing surgery or other invasive procedures. But the PillCam has changed all that. Invented by Israeli scientist Gavriel Iddan, it's a wireless, miniature camera that you swallow like an (oversize) pill. As it travels through your digestive tract, it takes about 50,000 shots of your intestines, lighting the way with six LEDs. It sends the pics to a data recorder you wear on a belt. No need to stay in the hospital—you go through your day while the PillCam snaps away, then it turns its photos into a video that'll give your doc a clear view of your insides. When the cam's work is done, it gets safely flushed away.

NERD ALERT: DO-GOOD GEAR

BEFORE THE PILLCAM came along, an endoscope, which is a long tube with a camera on the end of it, was the go-to for imaging your guts. But it couldn't easily reach the small intestine, which is between six and 26 feet (1.8–8 m) long. The PillCam gets through during normal digestion.

FUN FACT

Bioengineers are working on a new "robotic" pill that could replace insulin injections. It delivers a tiny injection inside your stomach, where you can't feel it.

FUN FACT

The ALICAM works just like a PillCam ... but for dogs!

DARKEST MATERIAL ON EARTH

VANTABLACK PAINTED ON CRINKLED FOIL

WHAT'S BLACKER THAN BLACK? The answer is a nanotube coating that accidentally became the darkest human-made material in the world. Absorbing almost all light (99.995 percent of it), it was created at MIT when engineers were experimenting with ways to form carbon nanotubes (CNTs). CNTs are microscopic hollow tubes made of carbon atoms, and they're really useful because of how well they conduct electricity. When the scientists were done, the tubes were surprisingly dark—so dark that the engineers decided to test them. It turned out that they soak up almost all light that hits them. When diamonds were coated with these nanotubes, they looked like flat, black cutouts! Though it appears velvety, the coating feels smooth because the nanotubes bend under your fingers. The nanotubes will be used to absorb light inside microscopes and telescopes, making the view clearer.

NERD ALERT: SCI-FI COOL

UNTIL RECENTLY, the blackest material on Earth was another nanotube coating called Vantablack. It absorbed a little less light than this new substance, but to human eyes, it's hard to tell the difference. In fact, when it coats crinkled foil, the foil looks like a hole because there's no reflection to show dimensions!

FUN FACT

Anish Kapoor is the only artist licensed to use Vantablack, and he used it to create what look like black holes.

FUN FACT

Artist Stuart Semple created Black 3.0, an ultra-dark matte paint, so that all artists can get the feeling of working with something similar to nanotubes.

NIGHT SKY? NOPE.
THIS IS A BUILDING SPRAY-PAINTED WITH VANTABLACK IN PYEONGCHANG, SOUTH KOREA.

HOT WHEELS

WHAT MIGHT YOUR FUTURE CAR BE LIKE? CHECK OUT THE CUTTING-EDGE CAR TECH BEING DEVELOPED NOW.

BRAIN-TO-VEHICLE

AUTOMAKERS HAVEN'T YET REVEALED whether it'll take a cap filled with sensors or maybe new tech we haven't imagined. But however it works, it'll track your neural activity to detect when you're about to brake, accelerate, or turn, and make that happen before your body catches up—so you can avoid accidents better than ever.

OUTSIDE AIRBAGS

THESE WILL EXPAND OUTSIDE YOUR CAR LIKE GIANT MARSHMALLOWS if the car detects a crash is about to happen. The airbags will slow the car to prevent the accident.

VIRTUAL VIEW

WHAT'S THAT OUTSIDE YOUR WINDOW?! Thanks to augmented reality, passengers will be able to zoom in for a better look, while drivers will call up directions that appear as holograms in their windshield, so they won't need to look away from the road.

TALKING CARS

EVEN IF YOU DON'T NOTICE another car about to run a red light, your car will stop just in time. That's thanks to "vehicle to vehicle communication." Silent messages are broadcast up to 10 times a second to "tell" other vehicles where your car is, where it's heading, and how fast it's going. Then it sends you an alert when it detects potential danger, including some you can't see.

SPEAKING OF STAYING CONNECTED, CHECK OUT HOW THE NERD OF NOTE ON THE NEXT PAGE IS HOOKING UP AN ENTIRE CONTINENT.

NERD OF NOTE:
JULIANA ROTICH

When she was a child, growing up in Kenya, Juliana first learned about astronaut Mae Jemison, the first Black woman to go to space. "That was so encouraging and inspirational to me," Juliana says. It helped her feel sure she could someday have a career in technology, too.

Juliana studied tech in college and took jobs in the field. Then, in 2008, a Kenyan election ended in violence. Juliana noticed that TV reports were different from what she and others witnessed on the streets. She realized that people were missing important news—news that could come only from the local people experiencing it. So she put her skills to work. She co-created Ushahidi, a web-based reporting system that crowdsourced data in real time.

Yet many people can't afford to connect to the internet in Africa, and even those who can, will lose connection during frequent power outages. To address this, Juliana co-created BRCK, a block-shaped charger and Wi-Fi hot spot in one. She also co-created Moja, which is ad-supported free internet—users don't have to pay for access.

Proudly calling herself an "African geek girl," Juliana hopes to inspire more tech creators. "I think, for girls, it's really important to show them that it is possible to dream big," says Juliana, "and if your dreams include some nerdy creation, that's fine."

"Technology still has the power to inspire, to create solutions, to create a sustainable living for anyone, anywhere."

JULIANA ROTICH, tech entrepreneur

FUTURE FARMING

HERE ARE SOME OF THE MODERN MARVELS THAT ARE CHANGING FARMING FOREVER.

BIG PULL

ORGANIC FARMING IS GREAT, except when pesticide-free crops get eaten by hungry bugs. How can you stop insects without using toxins? With a bug vacuum! This massive vac attaches to the back of a tractor and sucks bugs off even delicate plants.

LOOKING UP

AEROFARMS GROW INDOORS in stacked "farming racks," each with their own LED light source and misting system that delivers water, fertilizer, and oxygen. They grow 390 times more per square foot than a traditional farm and use up to 95 percent less water.

SERIOUS DRIVE

WHETHER FARMERS ARE WATCHING THEIR CROPS or are still sound asleep, these tractors keep rolling along. Using sensors similar to those in self-driving cars, self-driving tractors can plow fields day and night without a break, giving humans time for other work.

TECHNO-HOUSES

HOW ABOUT A GREENHOUSE that sends down a shade cloth when the sun gets too strong, opens a vent when it's too steamy, or sounds an alarm if the temperature creeps too high? Computer-controlled greenhouses can do all of this and also check nutrient levels so that just the right amount of food goes to each plant.

NOISE-CANCELING HEADPHONES

FUN FACT

When Bose designed noise-canceling headphones for the NFL, they had to work in freezing cold, wind, and rain—all the conditions players experience during practice!

EVER TRY TO LISTEN TO A MOVIE ON A PLANE, only to hear mostly jet engine noise? Enter: noise-canceling headphones. "Passive" models use lots of foam to blot out ambient sound, but foam blocks only about 20 dB (decibels), and jet engines blast you with up to 80 dBs inside the cabin. That's where "active" noise-canceling headphones come in. These literally cancel out sound waves by countering them with their own waves that are 180 degrees "out of phase" with the noisy ones. To do that, a microphone inside the headphones picks up surrounding noise. Electronics in the ear cup map the incoming sound wave, and generate a canceling wave. The headphones' speakers emit that canceling wave so you hear only the music you want to hear ... or sweet silence.

NERD ALERT: Everyday Egghead

IF YOU'VE HEARD OF BOSE ELECTRONICS, then you've also heard of the inventor of noise-canceling headphones, Dr. Amar Bose. The MIT prof was inspired during a 1978 flight from Zurich, Switzerland, to the U.S., when engine noise kept him from hearing through his headphones.

NOISE-CANCELING HEADPHONES CAN HELP BLOCK OUT THE SOUNDS OF A NOISY FLIGHT.

FUN FACT

Noise-canceling headphones can't protect your ears against sudden loud noises. Why not? There's a short delay before the electronic noise-masking signal kicks in.

NERDSVILLE CENTRAL: CERN

A COSMIC CURIOSITY

WHAT CAN THE BIGGEST MACHINE IN THE WORLD DO? The Large Hadron Collider (LHC), housed at CERN at the Swiss and French border, can smash subatomic particles together to re-create the big bang in miniature. This allows physicists to peek at the kinds of particles that were produced when our universe began, and better understand how our world works. Talk about studying history! You can walk through CERN's permanent exhibitions to learn what the LHC has found so far, explore exhibits about the smallest particles known to science, and more. While you can't step inside the LHC itself, you will stand right on top of it, see models of it, and see the parts used to create it. Plus, you can say once and for all that you got close to the collider—which many people worried would wreak havoc when it was first switched on—and weren't sucked into a black hole. Probably.

EMERGENCY

1 Press the "emergency button." It's red, it's inviting, and you're allowed to push it, so give it a try!

2 Ask about the ice cream. Because the LHC needs extreme cooling to work, it uses a cryogenic (meaning very low temperatures) system. CERN sometimes holds special events during which they make ice cream using cryogenics, and you can sample it.

TIPS TO NERD OUT AT CERN:

3 Take a tour. They're free and you'll get to go inside the operations center of a CERN experiment!

4 Cross the road to the Globe of Science and Innovation. The huge dome is a landmark, with a particle physics light show inside.

ELEKTRO AND SPARKO

PICTURE A SEVEN-FOOT (2.1-M)-TALL ROBOT that responds to voice commands, has a 700-word vocabulary, and can perform 26 actions, including blowing up balloons and moving each finger separately. Throw in a robo-dog that sits on cue, begs, and wags its tail, and you might not guess this pair of bots was created in the 1930s. Elektro and Sparko were built by Westinghouse Electric to display at the 1939 World's Fair. To command Elektro, you'd speak into a microphone. Your words' vibrations were turned into electrical pulses that triggered a lightbulb flash. That flash then traveled to an "electric eye," through a telephone relay, and then back to Elektro to activate him. The number of syllables in each command mattered—a three-one-two pattern like "Will you come ... down ... front, please?" would start the action. Speaking of voices, Elektro's came from records, not artificial intelligence.

SPARKO

NERD ALERT: VINTAGE VISIONARY

ELEKTRO'S FINGERS COULD MOVE, thanks to wire "tendons" connected to a motor. Honda's modern ASIMO robot moves its fingers, too, but in a whole new way. A motor in ASIMO's chest links to each finger and uses hydraulic fluid like that used in car brakes. The fluid's pressure gives Asimo a good grip. Each fingertip also has sensors to "feel" the object, so it knows how strongly or gently to grab things.

FUN FACT

Parts of Elektro were assembled from coffeepots, waffle makers, and irons.

FUN FACT

Elektro had a window in his middle so that audiences could see there wasn't a person operating him from inside.

ELEKTRO "CONDUCTS" AT THE 1939 WORLD'S FAIR.

ROBOT LOVE MATCHES

WHO SAYS MACHINES CAN'T FALL IN LOVE? OK, THEY CAN'T. BUT IF THEY COULD ... CHECK OUT THESE MADE-UP ROBOT LOVE MATCHES.

ELEKTRO AND MARS ROVER CURIOSITY

NICKNAME: "Elekosity"

HOW THEY MET: After Elektro's entertainment career, he wanted to do something meaningful—like interplanetary travel. Curiosity was magnetically attracted.

FAVORITE THINGS TO DO: Exploring with the Mars rover makes Elektro feel futuristic again, and Curiosity likes hearing about the good old days.

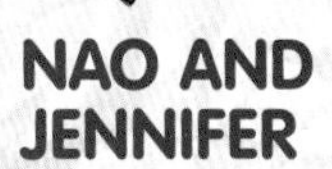

NAO AND JENNIFER

NICKNAME: "NAO-ifer"

HOW THEY MET: NAO the soccer-playing bot was getting a ball when it spotted sports bot Jennifer playing ice hockey. Both sensed an instant win.

FAVORITE THINGS TO DO: Perform in the Olympics—all the categories at the same time.

ROCKI AND QOOBO

NICKNAME: "Roobo"

HOW THEY MET: Rocki the cat toy robot was cruising by when it sensed Qoobo the tail-wagging cushion. Rocki tried running, giving out treats, and showing off its laser-pointer to get Qoobo's attention. But refreshingly, Qoobo was all cuddles and no games.

FAVORITE THINGS TO DO: Rocki's antics make Qoobo wag its tail, and Qoobo's chill attitude helps Rocki power down.

ZERO GRAVITY TRAINING

FLOATING AROUND ISN'T ALL FUN AND GAMES. BEFORE BLASTING INTO SPACE, NASA ASTRONAUTS MUST GET READY FOR LIFE IN ZERO GRAVITY. HERE'S SOME OF THE TECH THEY USE TO DO IT.

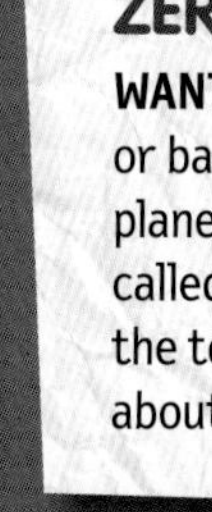

ZERO-G PLANES

WANT TO DO SOMERSAULTS IN THE AIR, or backflips off the wall? All it takes is a plane to fly you in an ultra-steep arc called a parabola. As the plane reaches the top of the parabola, you float for about 25 seconds.

NEUTRAL BUOYANCY LAB

IT'S LIKE A SWIMMING POOL ... with 6.2 million gallons (23.5 million L) of water, a 40-foot (12-m) depth, and a complete replica of the ISS in it. Astronauts jump in wearing full gear and practice spacewalks underwater. It feels a lot like space ... with bubbles.

POGO (PARTIAL GRAVITY SIMULATOR)

THIS BODY HARNESS AND RIG SEND YOU FLYING like a weightless astronaut. While you're soaring, you'll practice working on model space equipment and controlling your motion by grabbing on to whatever you can.

PRECISION AIR BEARING FLOOR (PABF)

HOW DO YOU MOVE A GIANT OBJECT WHEN IT'S FLOATING? Carefully. One touch, and a piece of equipment could fly out of control. You can practice moving frictionless items with this floor, which works like a massive air hockey table. Air shoots through tiny holes, so items slip-slide the way they would in space.

AIR HOCKEY TABLE

ARTIFICIAL FLAVORS

FUN FACT

Most people can't ID the flavor of Juicy Fruit gum, but some suspect it's isoamyl acetate—a chemical found in jackfruit.

THESE MIGHT GET A BAD RAP—AFTER ALL, if you're not a food chemist, you probably don't recognize their ingredients. Well, tune in your taste buds for a lesson. Natural flavors are made of hundreds of natural chemicals. Artificial flavors are made from chemicals, too—often, identical ones to those in natural flavors. The difference is their source. Natural flavors come from stuff you can eat. Artificial flavors come from stuff you wouldn't ordinarily eat, like paper or petroleum. So why bother faking it? Reproducing flavors can be a lot cheaper than using the originals. Vanillin, the taste and smell of vanilla, comes from a specific orchid, making it pricey. One researcher found a way to get vanillin from cow poop. Um ... sweet?

JACKFRUIT

NERD ALERT: Everyday Egghead

WHEN A LABEL LISTS "NATURAL FLAVORS," it means the flavors were still formulated in a lab, but using edible components—though not necessarily the same ones they're imitating. To make "natural" fruity flavors, for instance, food chemists start with esters, chemicals that come from fruits, yeast fermentation, or flowers. So your "natural" watermelon lollipop might be flavored with African violets.

FUN FACT

Up to 80 percent of what you taste comes from smell, so artificial flavors usually include scents to make them seem even tastier.

TURN THE PAGE TO GET A SENSE OF SOMEONE WHO'S USING SCENT IN AROM-AZING WAYS.

NERD OF NOTE:
ANI LIU

Growing up, Ani Liu wanted to be an artist, but her parents hoped she'd study something more "practical," like math or science. So she became an artist and a technologist, and now she finds new ways for people to sense the world!

In college, Ani learned about microbiomes, which include "all the organisms that you don't see but that live on or in you." She was fascinated to discover that more than half of the cells in your body aren't your own. "The idea that other organisms that live inside me could contribute to my identity was really interesting to me," Ani says. So she started making microbial "portraits"—bacterial cultures of the microscopic organisms from her own body. And she kept going from there.

Since then, Ani's created many other projects. She found a way to distill people's scents into a kind of human perfume, to create plants that smell like familiar people. And she's created Biota Beats, a record player that plays different sounds based on your body's bacteria.

Ani is excited to show people how science and art can work well together. "I feel artists have been making self-portraits for thousands of years," she says. "So, what does a self-portrait that includes this new scientific knowledge look like?"

"I feel it's really important to have a more humane understanding of technology."

ANI LIU, artist and technologist

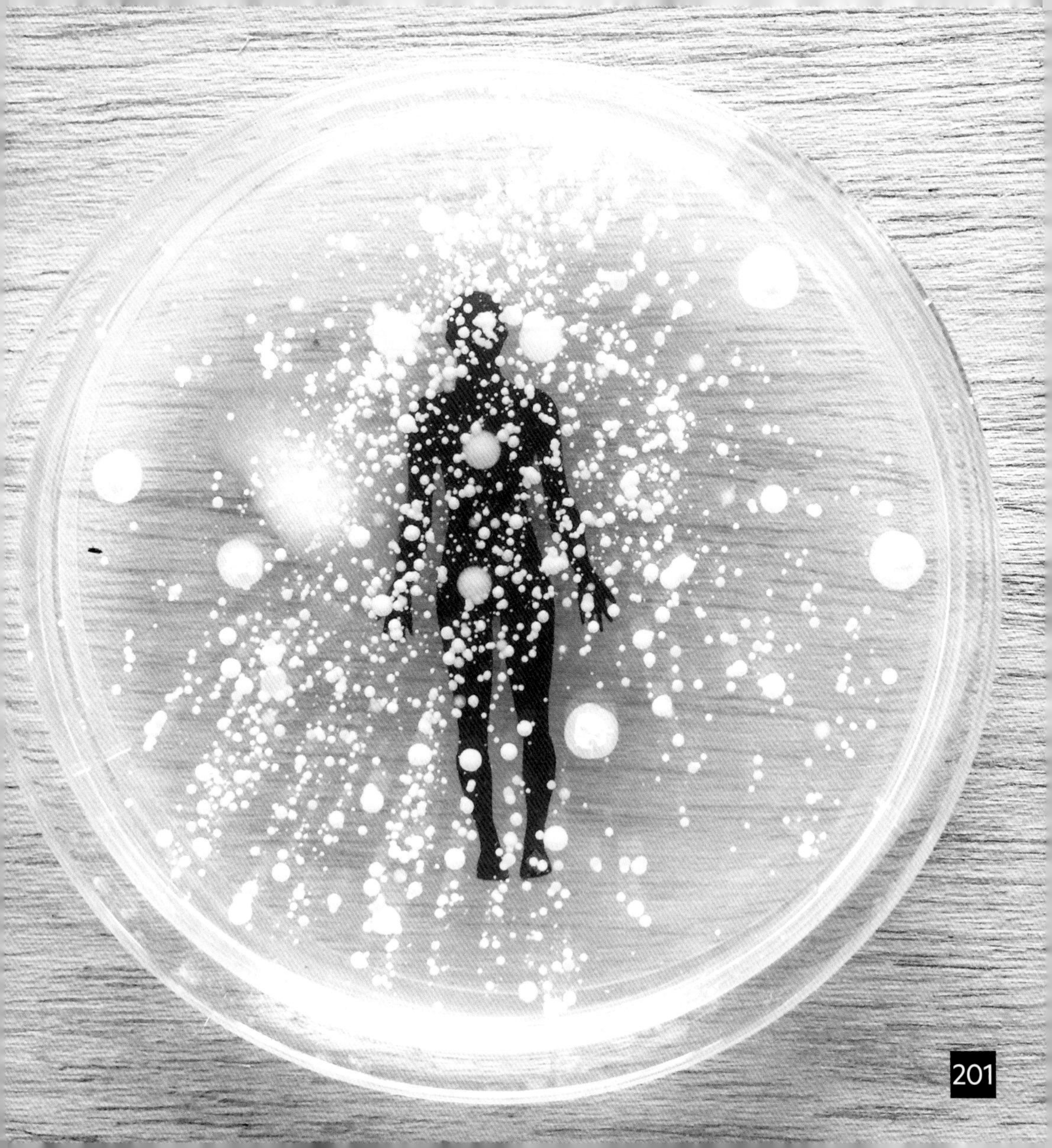

INTERNET OF ANIMALS

ISS

A SINGLE MOUNTAIN LION CALLED P-22 pads through the Santa Monica Mountains of Los Angeles, U.S.A. He's kind of a big deal, because he's the only known mountain lion in the area. Fortunately, scientists can track exactly where he is and help look out for his safety, thanks to the Internet of Animals (IoA), a tracking system for animals all over the planet. Scientists and environmentalists use the IoA to keep track of populations and migration patterns, and how climate and environmental changes affect animals. Trackers also alert rangers when they detect gunfire, so poachers can be caught. They might even help predict natural disasters by picking up when animals flee areas or act unusually, potentially signaling earthquakes or floods. And thanks to mini insect trackers, scientists can recognize insect population booms and help farmers prepare.

FUN FACT

The ICARUS animal tracking system uses the International Space Station as its satellite.

NERD ALERT: DO-GOOD GEAR

EARLIER ANIMAL TRACKERS TRANSMITTED SIGNALS to radio antennas. Today's GPS trackers receive instead of transmit. Like your cell phone, they get signals from satellites. It might seem like a small change, but modern trackers are making a big difference in conservation around the world.

FUN FACT

You can help track animals with the Animal Tracker app. It shows where tagged animals are in your area, so when you spot one, you can send comments on their real-life behavior.

TOILET TECH

SOME OF US DO OUR THINKING ON THE TOILET. OTHERS THINK ABOUT TOILETS—SPECIFICALLY, HOW TO MAKE THEM BETTER. HERE ARE SOME OF THE BIGGEST BREAKTHROUGHS.

THE BIOCYCLE

COULD YOUR POOP BE WORTH CASH? The BioCycle might make it valuable, thanks to larvae from the black soldier fly. Bugs turn waste into dollars by breaking it down into cash-worthy oil, fuel, and animal feed. And this means less pressure on landfills and sewer systems.

NANO MEMBRANE TOILET

IN SOME PARTS OF THE WORLD, water is so scarce that flushing with H_2O is out of the question. The Nano Membrane toilet vaporizes urine, pushes it through a membrane that removes pathogens, then collects the water for washing or irrigation. Poop is burned, creating energy to power the system and enough extra to charge your cell phone.

SPACE TOILET

WHERE DO ISS ASTRONAUTS DO THEIR BUSINESS? Behind a curtained area, there's a little vacuum that they pee into, and a teeny-weeny seat for doing number two. They just have to remember their seatbelt before they begin—this is zero gravity, after all. Thing is, sometimes, things go wrong and, uh, stuff floats out. If they spot an escapee, astronauts say the code "brown trout!" and whoever did it has to go catch it. Awkward.

PEE POWER

URINE AS FUEL? WHY NOT? This kind of microbial fuel cell (MFC) tech works by converting elements of pee into power through a series of batteries. And it produces enough energy to light a bathroom! It could also be used to charge your devices or run loo appliances, such as electric towel dispensers.

FOLD-SCOPE

MICROSCOPES ARE EXPENSIVE. In lots of places, would-be science nerds don't get a chance to use the instruments because their schools can't afford them. That's what got Stanford University Ph.D. student Jim Cybulski and his professor, Manu Prakash, thinking: Could they build a rugged microscope that costs less than a dollar? They were sketching ideas on paper when they had an idea: paper! Foldscope, a microscope made almost entirely of paper, was born. The pocket-size device can be assembled from a flat sheet in less than 10 minutes, gives more than 400 times magnification, and costs less than a dollar in parts. You can drop or step on it without breaking it, and it even has an LED that runs on a button battery.

A STUDENT PEEKS THROUGH A FOLDSCOPE.

NERD ALERT: DO-GOOD GEAR

BECAUSE FOLDSCOPES ARE SO INEXPENSIVE and easy to tote around, they're being used for more than lab work. They've found counterfeit money, tracked toxic blooms, and even identified microscopic eggs before they could hatch into crop-eating pests.

FUN FACT

If you put a light source behind a Foldscope, it turns into a projector.

INFINITE ENVIRONMENTS

IN MINECRAFT, NO TWO SCENES ARE ALIKE. In fact, you could literally play forever and not see the same landscape twice. This is called procedural generation. The idea is that a computer takes an algorithm—that is, a set of problem-solving rules—and follows its instructions to put certain building blocks in specific areas of the screen. More specifically, it starts with "noise maps," which are maps of textures. Using a noise map, the computer puts land textures and colors on the bottom of the screen and sky textures and colors on the top. Then it generates another noise map on top of the first, which adds details like valleys and hills, followed by another with trees and clouds. When it's done, it looks like a landscape someone had to think about, but it's really just a bunch of random calculations. Either way, it means the worlds you get will be unlike any others before or since.

A NOISE MAP

FUN FACT

Minecraft's noise maps are called "Perlin noise," and they were first developed by Ken Perlin for the original *Tron* movie.

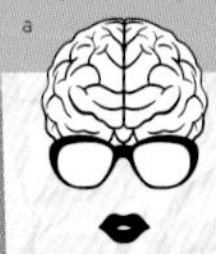

NERD ALERT: SUPER SMARTS

SOME OF THE WEIRD VOICES OF MINECRAFT'S CREATURES come from familiar things. The Enderman's language is mostly English words and phrases played backward, and the Ghast's whine is a cat being woken from a nap.

FUN FACT

Minecraft was once called Minecraft: Order of the Stone.

Unfortunately, most things—like this book—aren't infinite. *Fin.*

INDEX

Boldface indicates illustrations.

A

B

C

INDEX

CONTINUED

PHOTO CREDITS

AS: Adobe Stock; ASP: Alamy Stock Photo; GI: Getty Images; NGIC: National Geographic Image Collection; SS: Shutterstock

Cover (background), chekart/SS; (UP LE), Sergey Nivens/AS; (LO), Peter Bollinger; spine: tulpahn/AS; back cover (background), chekart/SS; (LE), chesky/AS; (RT), Natalia/AS; 1 (Throughout, circuit background), chekart/SS; 1 (Throughout, robot head icons), oxinoxi/AS; 1 (Throughout, bionic eye), riefas/AS; 1 (Throughout, 8-bit game icons), tulpahn/AS; 1 (Throughout, cloud technology icon), fahmi/AS; 1 (Throughout, monkey robot head), Annas Alam Yahya/SS; 1 (Throughout, graphing paper background), nicemonkey/SS; 1 (Throughout, retro hats), Chereliss/SS; 2-3, (background), chekart/SS; (UP LE), Sergey Nivens/AS; (LO), Peter Bollinger; 4-5, chesky/AS; 4 (UP LE), Visual China Group/GI; 4 (UP RT), boule1301/AS; 5 (Arcade games monsters), ArcadeImages/ASP; 5 (Kiki Robot), AP Photo/John Locher; 6 (Throughout, triangle with plaid pattern), Anya D/SS; 6, Sergii/AS; 7 (Throughout, old paper), Thammasak_Chuenchom/Thinkstock Images; 7 (Throughout, mustache), Anna Marynenko/SS; 7, Lawrence Livermore National Laboratory; 8-9 (Throughout, benzene rings background), BQ-Studio.ru/AS; 8, Joni Hanebutt/SS; 8 (Throughout, brains), kaiwut niponkaew/SS; 8 (Throughout, eyeglasses), Fosin/SS; 8 (Throughout, lips), Vlada Young/SS; 9, Prostock-studio/AS; 10 (LE), NASA/JPL-Caltech; 10 (RT), Alexander Gos'kov/AS; 11 (LE), vitaly tiagunov/AS; 11 (RT), Festo AG & Co. KG; 12-13 (MONONOFU), Richard Atrero de Guzman/Nippon News/AFLO/ASP; 12 (Throughout, graduation cap), Creativee Icon/SS; 13 (Xingzhe), VCG/GI; 13 (Alpha 1S), AP Photo/Liu haitian - Imagine china; 13 (Forpheus), Kim Kyung-Hoon/Reuters; 13 (Tradinno), CB2/ZOB/WENN/Newscom; 14 (UP), NASA Photo/ASP; 14 (Throughout, knitted hats), Reinke Fox/SS; 14 (LO), AP Photo/Kevin Wolf; 15, NASA; 16, Prostock-studio/SS; 17, Artur Debat/GI; 17 (Throughout, grunge strip), monbibi/SS; 18 (UP), Amber Zoe Yang; 18 (LO), NASA/JSC; 19, Andrey VP/SS; 20, Steve Winter/NGIC; 21, John Downer Productions/Nature Picture Library; 22 (labyrinth), Igor Osypenko/AS; 22 (chess), pio3/SS; 22 (barrels), ArcadeImages/ASP; 23 (arcade games monsters), ArcadeImages/ASP; 23 (Mario), ArcadeImages/ASP; 23 (red hammers), ArcadeImages/ASP; 23 (red circle), Arcade Images/ASP; 23 (beetle), panor156/SS; 23 (praying mantis), dwi/AS; 23 (spaceships), koya979/SS; 23 (gorilla), ArcadeImages/ASP; 23 (player and arcade game cabinet), danielegay/AS; 24-25 (ALL), NASA; 26 (Throughout, cracks), Vector FX/AS; 26 (seismogram), Vector FX/SS; 27 (termites), Pan Xunbin/SS; 27 (UP LE), Zeynur Babayev/Dreamstime; 27 (UP RT), Todd Korol/Bloomberg/GI; 27 (LO LE), Ken Wilson-Max/ASP; 28 (Throughout, binary background), Tartila/AS; 28, Wiskerke/ASP; 29, MacRebur; 29 (inset), Regis Duvignau/Reuters; 30, The Ocean Cleanup Handout/EPA-EFE/SS; 31, The Ocean Cleanup/SS; 31 (rubber duckies), by-studio/AS; 32-33 (Throughout, starburst background), dynamic/SS; 32, TCD/Prod.DB/ASP; 33 (UP), Clay Enos/Warner Bros/Kobal/SS; 33 (LO), Pictorial Press Ltd/ASP; 33 (germs), sveta/AS; 34 (UP), SeanPavonePhoto/AS; 34 (bamboo shoots), cameilia/SS; 35, Alexandre Rotenberg/SS; 36, Sergey Novikov/AS; 37, lenscap67/GI; 38-39, Thierry Thorel/NurPhoto/GI; 40 (dragonfly), Ivan/AS; 40 (RT), Charly Triballeau/GI; 41, Sarcos Corp.; 41 (crab), Artinun/AS; 42-43 (Akihabara), Lucas Vallecillos/ASP; 42 (Throughout, map), twenty1studio/SS; 43 (LO), Tatiana Andrianova/AS; 44-45, Stocktrek/GI; 44, WENN Rights Ltd/ASP; 46, Aline Lessner/Stringer/AFP/GI; 47, AndriyShevchuk/SS; 48 (UP RT), voltan/AS; 48 (LO LE), supamas/AS; 49 (UP RT), onairjiw/AS; 49 (LO LE), Solar Foods; 50 (UP LE), scharfsinn86/AS; 50 (LO RT), Akarat Phasura/AS; 51, Mike Mareen/AS; 52, Oli Scarff/AFP/GI; 53 (UP LE), Bruce R. Bennett/The Palm Beach Post/ZUMA Press/ASP; 53 (CTR RT), Empatica Inc.; 53 (LO LE), pixelrobot/AS; 54-55 (ALL), Patrick McCallum; 56-57, VCG/GI; 57 (MONONOFU), Richard Atrero de Guzman/Nippon News/AFLO/ASP; 57 (Alpha 1S), AP Photo/Liu haitian - Imaginechina; 57 (Forpheus), Kim Kyung-Hoon/Reuters; 57 (Tradinno), CB2/ZOB/WENN/Newscom; 59, Lenscap Photography/SS; 60-61 (paw prints), MicroOne/SS; 60, Whistle Labs, Inc.; 61 (UP RT), Automated Pet Care Products, Inc.; 61 (CTR LE), AP Photo/Ross D Franklin; 61 (LO RT), Petcube, Inc; 62 (mosquitos), Potapov Alexander/SS; 62 (Thermacell device), Thermacell Repellents, Inc.; 63, Thermacell Repellents, Inc.; 64 (globe), gst/SS; 64 (car), GioGioGio/AS; 64 (plane), Sfio Cracho/AS; 64 (question marks), 4zevar/AS; 65 (robot), phonlamaiphoto/AS; 65 (passenger drone), chesky/AS; 65 (child with jet pack), Sunny studio/AS; 66-67 (pizza), victoria pineapple/SS; 66, Nuro/Kroger/Cover Images/Newscom; 67 (UP LE), Valda Kalnina/EPA-EFE/SS; 67 (RT), John D. Ivanko/ASP; 68 (LE), B Christopher/Alamy Stock; 68 (RT), robuart/SS; 69, Kevin Britland/ASP; 70-71, andrey_l/SS; 71 (black clock), lunglee/AS; 71 (red clock), dimedrol68/AS; 72, Great Wight Productions Pty Ltd and Earthship Productions, Inc./NGIC; 73, Mark Thiessen/NGIC; 74 (LE), Yevhen Prozhyrko/SS; 74 (RT), Paul J. Richards/AFP/GI; 75, Olivier Douliery/ABACAUSA/Newscom; 76-77, Toni L. Sandys/The Washington Post/GI; 76-77 (Shuriken), quatrovio/SS; 77 (James Bond), Andreas Rentz/WireImage/GI; 78, NASA; 79, Paolo Nespoli/ESA/NASA/GI; 80, Verge Aero; 81 (UP LE), David Gray/Bloomberg/GI; 81 (CTR RT), AP Photo/Mark Frydrych/NOAA Fisheries; 81 (LO LE), Peter Parks/AFP/GI; 82 (UP), chesky/AS; 82 (LO), NFT, Inc.; 83, chesky/AS; 84 (LE), sivvector/AS; 84 (RT), Tramino/GI; 85, Yoshio Tsunoda/

Nippon News/Aflo Co. Ltd./ASP; 86 (UP), Fotoscenen; 86 (LO), Fotoscenen; 87, ZCHD/Samuel Westergren/Newscom; 88 (UP), Richard Olsenius/NGIC; 88 (LO), warpaintcobra/AS; 89, Jeff J. Mitchell UK/Reuters; 90, Brian J. Skerry/NGIC; 91, anyamuse/SS; 92-93 (blue tablet), Paper Street Design/SS; 92, Eric Isselée/AS; 93 (UP), Tony Campbell/AS; 93 (LO), Ivonne Wierink/AS; 94-95 (all), Christian Miller; 96-97, AP Photo/Liu haitian - Imaginechina; 97 (MONONOFU), Richard Atrero de Guzman/Nippon News/AFLO/ASP; 97 (Xingzhe), VCG/GI; 97 (Forpheus), Kim Kyung-Hoon/Reuters; 97 (Tradinno), CB2/ZOB/WENN/Newscom; 98, Pictorial Parade/ Archive Photos/GI; 99, Andrey Popov/AS; 100-101, RenataA photography/SS; 103, rh2010/AS; 104-105, Jack Taylor/Stringer/ GI; 104 (arcade games monsters), ArcadeImages/ASP; 105 (airplane icon), Bowrann/SS; 106 (Wi-Fi icon), Web-Design/SS; 106 (UP LE), phonlamaiphoto/AS; 106 (LO RT), AP Photo; 107, KenSoftTH/SS; 108-109, Shutter_M/SS; 111, Dulwich Picture Gallery/Bridgeman; 111 (inset), AP Photo; 112-113 (Z6 robot), Robugtix Ltd.; 112 (blue morpho butterflies), boule1301/AS; 112 (LO LE), Visual China Group/GI; 113 (bees), Daniel Prudek/AS; 113 (UP), Thierry Falise/LightRocket/GI; 113 (cockroaches), James Steidl/AS; 113 (LO), Prof. Robert J. Full; 114, Andrew Brookes/ Image Source/GI; 115, ktsdesign/AS; 116 (monkey robot head), Annas Alam Yahya/SS; 116 (waving robot), phonlamaiphoto/AS; 116 (robot hugging a heart), Blueprint Characters/SS; 116 (floating robot), ogro/AS; 117 (Tamagotchi Toys), Business Wire/GI; 117 (Robear), Riken/SS; 117 (Cheetah robot), Photographer/GI; 118, Andrey Armyagov/AS; 119, Monopoly919/AS; 120-121 (all), Dr. Sarah Parcak; 122 (Distant Galaxies), NASA; 122 (Helix Nebula), NASA/ NGIC; 123, marcel/AS; 124-125 (knitting icon), ArtMari/SS; 124, David Paul Morris/Bloomberg/GI; 125 (UP RT), divedog/AS; 125 (CTR RT), Qnature UG; 125 (LO LE), Orange Fiber S.R.L; 126 (UP LE), Andrey Popov/AS; 126 (LO RT), zapp2photo/AS; 127, tilialucida/AS; 128 (LO RT), Queensland University of Technology; 128 (UP RT), BillionPhotos.com/AS; 129, pressmaster/AS; 130, George W. Bailey/SS; 131 (UP), MivPiv/GI; 131 (tape), Yasemeen.P/SS; 131 (LO RT), Katherine Frey/The Washington Post/GI; 131 (LO LE), Deb Lindsey/The Washington Post/GI; 132, diy13/AS Photo; 133, Thongchai Rujiralai/Dreamstime; 134, Farlap/ASP; 135, Ruslan Kalnitsky/SS; 136 (CTR LE), Bettina Strenske/ASP; 136 (LO), Dmitrii Morozov; 137 (UP LE), Eric Standley; 137 (CTR RT), Minimaforms; 137 (LO LE), Dmitrii Morozov; 138-139, Kim Kyung-Hoon/Reuters; 139 (MONONOFU), Richard Atrero de Guzman/Nippon News/AFLO/ ASP; 139 (Xingzhe), VCG/GI; 139 (Alpha 1S), AP Photo/Liu haitian - Imaginechina; 139 (Tradinno), CB2/ZOB/WENN/Newscom; 140, Alexander Zhiltsov/Dreamstime; 141 (UP LE), Meredith Ogle; 141 (CTR RT), JLM Innovation GmbH; 141 (LO LE), Prof. Liangfang Zhang; 142-143, GL Archive/ASP; 144, AP Photo/Chitose Suzuki; 145, Henrik5000/GI; 146-147 (all), Images courtesy of Mammutlab, Wood-skin, and Self- Assembly Lab of MIT; 147 (bunny), brulove/AS; 148-149 (cartoon bubble), Kapitosh/SS; 148, frog/AS; 149 (UP), Pally/Alamy Stock; 149 (LO), Kraig Biocraft Laboratories, Inc.; 149 (inset), Kraig Biocraft Laboratories, Inc.; 150-151 (test tube), AllNikArt/AS; 150 (LE), Franco Banfi/Nature Picture Library; 150 (RT), Derek Holzapfel/Dreamstime; 151 (LO), Dr. Pablo Berea Núñez; 151 (UP), Matthijs Kuijpers/ASP; 152, Unimedia/SS; 153, Jeff Gilbert/ASP; 154, Bettmann/GI; 155, Xavier Meal/Caters News Agency; 156, Stephen Barnes/Science/ASP; 157, Kenneth Garrett/NGIC; 158-159, JP5\ZOB/WENN/Newscom; 158, Anton Gvozdikov/SS; 160 (UP LE), NASA/JPL-Caltech; 160 (LO RT), NASA/JPL-Caltech/MSSS; 161, NASA/JPL-Caltech; 162 (LE), Radomir Rezny/SS; 162 (RT), Solarisys/AS; 163 (UP LE), Jason Pennypacker; 163 (RT), Haruyoshi Yamaguchi/Reuters; 164, University of New Hampshire/Gado/GI; 165, Z1022 Patrick Pleul/ Deutsch Presse Agentur/Newscom; 166, Kazuhiro Nogi/AFP/GI; 167 (UP), AP Photo/John Locher; 167 (CTR RT), Yukai Engineering, Inc.; 167 (LO LE), Tomohiro Ohsumi/Stringer/GI; 168, Wachiwit/SS; 169, David Espejo/GI; 170, Taro Karibe/Stringer/GI; 171, Christophe Archambault/AFP/GI; 172 (TIE Fighters), Lucasfilm; 172 (X-Wing), Lucasfilm; 172 (Millennium Falcon), Lucasfilm; 172 (Drilling tool), Prokhorovich/SS; 173 (R2-D2), Lucasfilm; 173 (toaster), Prokhorovich/SS; 173 (game controller), Prokhorovich/ SS; 173 (C-3PO), jpgfactory/GI; 173 (watch), Netkoff/SS; 173 (webcam), Netkoff/SS; 173 (writing pad), Netkoff/SS; 173 (Doraemon and Nobita), Dream Story/SS; 174-175, CB2/ZOB/ WENN/Newscom; 175 (MONONOFU), Richard Atrero de Guzman/ Nippon News/AFLO/ASP; 175 (Xingzhe), VCG/GI; 175 (Alpha 1S), AP Photo/Liu haitian - Imaginechina; 175 (Forpheus), Kim Kyung-Hoon/Reuters; 176, AP Photo/Tamara Lush; 177 (UP RT), Splash News/Sea Turtle Inc./Newscom; 177 (CTR LE), Phoenix Herpetological Sanctuary; 177 (LO RT), Caters News Agency; 178, imageBROKER/ASP; 179, David Bleeker Photography/Alamy Stock; 180, ZJAN/Surrey NanoSystems/Newscom; 181, ITAR-TASS News Agency/ASP; 182 (all), Kim Kyung-Hoon/Reuters; 183 (CTR RT), Coneyl Jay/GI; 183 (LO LE), Karsten Neglia/SS; 184 (UP), Waldo Swiegers/Bloomberg/GI; 184 (LO), World History Archive/ASP; 185, Brian Otieno/Scopio; 186 (LE), Eric L Tollstam/SS; 186 (RT), Mischa Keijser/GI; 187 (UP), Jonas Gratzer/LightRocket/GI; 187 (CTR), Suwin/SS; 188, Lightfield Studios/AS; 189, Eternity in an Instant/ GI; 190-191, xenotar/GI; 190, Meawpong/AS; 191, Ronald Patrick/ GI; 192, Bettmann/GI; 193, Hulton Archive/Stringer/GI; 194 (LE), Bettmann/GI; 194 (RT), freestyle_images/AS; 195 (UP CTR), Sebastian Willnow/picture alliance/GI; 195 (UP RT), Chris Iverach-Brereton; 195 (Qoobo), Yukai Engineering, Inc.; 195 (Rocki), Rocki/ Cover-Images/Newscom; 196 (LE), NASA/Bill Stafford; 197 (UP), NASA Image Collection/ASP; 197 (LO RT), Nina/AS; 197 (LO LE), Benoit Tessier/Reuters/Alamy; 198, nakornchaiyajina/AS; 199, New Africa/AS; 200-201 (all), Ani Liu; 202, NASA; 203, Steve Winter/NGIC; 204-205 (icons), nikiteev_konstantin/SS; 204, Autumn Sky Photography/SS; 205 (UP LE), Alison Parker; 205 (CTR RT), James Blair/NASA/JSC; 205 (LO LE), Ladanifer/SS; 206-207 (ALL), Foldscope Instruments, Inc.; 208, John David Martin; 209, Georg Wendt/dpa/ASP

To you, for being brave enough to show what you know.
Keep learning! —JKA

Since 1888, the National Geographic Society has funded more than 14,000 research, conservation, education, and storytelling projects around the world. National Geographic Partners distributes a portion of the funds it receives from your purchase to National Geographic Society to support programs including the conservation of animals and their habitats. To learn more, visit natgeo.com/info.

For more information, visit nationalgeographic.com, call 1-877-873-6846, or write to the following address:

National Geographic Partners, LLC
1145 17th Street NW
Washington, DC 20036-4688 U.S.A.

For librarians and teachers: nationalgeographic.com/books/librarians-and-educators

More for kids from National Geographic: natgeokids.com

For rights or permissions inquiries, please contact National Geographic Books Subsidiary Rights: bookrights@natgeo.com

Designed by Brett Challos

Library of Congress Cataloging-in-Publication Data

Names: Kiffel-Alcheh, Jamie, author.
Title: Nerdlet Tech/by Jamie Kiffel-Alcheh.
Description: Washington, D.C.: National Geographic Kids, 2022. | Series: NERDlet | Includes index. | Audience: Ages 8-12 | Audience: Grades 4-6
Identifiers: LCCN 2021019783 | ISBN 9781426373589 (trade paperback) | ISBN 9781426373596 (library binding)
Subjects: LCSH: Technology--Miscellanea--Juvenile literature. | Inventions--Miscellanea--Juvenile literature.
Classification: LCC T48 .K53 2022 | DDC 600--dc23
LC record available at https://lccn.loc.gov/2021019783

The publisher would like to thank the nerds who made this book possible: Kathryn Williams, project editor; Sarah J. Mock, senior photo editor; Danny Meldung, photo editor; Robin Palmer, fact-checker; Ebonye Wilkins, reviewer; and Anne LeongSon and Gus Tello, associate designers.

Printed in China
22/PPS/1

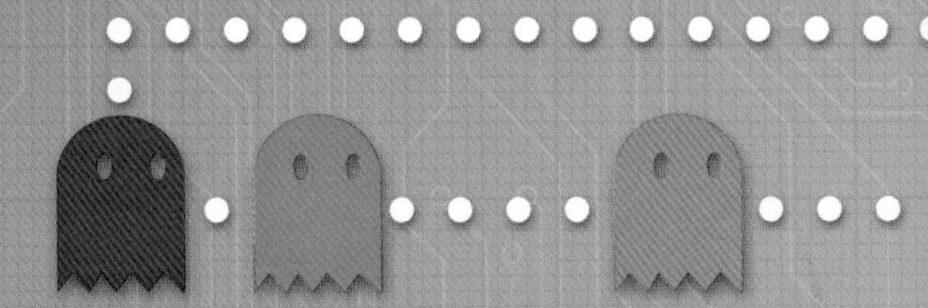